PRAISE FOR *MYCORRHIZAL PLANET*

"Let's make soil great again. Michael Phillips and *Mycorrhizal Planet* have a plan. This book focuses on the tangible things you can do on the land you love to make it healthier and more productive. In forty years doing USDA research and producing mycorrhizal fungi, I have read and reviewed thousands of mycorrhizal articles. Michael Phillips gets it. Finally we have a mycorrhizal book that is entertaining, practical, and vibrant. We truly live on a mycorrhizal planet, and people who read this book will emerge with a profound understanding of how these little creatures shape our earth and our future."

— **Dr. Mike Amaranthus**, founder, Mycorrhizal Applications

"*Mycorrhizal Planet* offers fascinating science and practical ideas for gardeners, farmers, foresters — for everyone, in fact. Learning how we can work with beneficial soil fungi is deeply relevant, not only to support optimal plant health and nutrition but as part of a lasting climate change solution."

— **Eric Toensmeier**, author of
The Carbon Farming Solution

"Michael Phillips is an emissary from the fungal realm, and he's here to tell us, through both study and practice, how our partnership with fungi is not only crucial but how it can be carried out practically on our homesteads and farms."

— **Ben Falk**, author of
The Resilient Farm and Homestead

"*Mycorrhizal Planet* isn't just a book about wild-running fungi. It covers in great detail all the benefits, scientific research, and technical information known about mycorrhizae. It also outlines methods of how to manage soils with the use of organic fertilizers, crops grown, and proper tillage to get the biology to flourish — including mycorrhizae. Because if a grower knows *why*, he or she will teach themselves *how*."

— **Gary Zimmer**, founder, Midwestern BioAg;
author of *The Biological Farmer*

"In his new book, *Mycorrhizal Planet*, Michael Phillips weaves his own web of astounding connections regarding what holds this earth together. Not since Paul Stamets's pioneering inquiry, *Mycelium Running*, have we been blessed by such a synthesis that tells how symbiotic fungi are the true and most trustworthy stewards of this planet."

— **Gary Paul Nabhan**, author of
Growing Food in a Hotter, Drier Land

"Fungi are not just decomposers, they are composers of soil and orchestrators of soil biodiversity. *Mycorrhizal Planet* pays tribute to the small and unseen, the uncredited collaborations beneath our feet, and Michael Phillips leads the tour underground for everyone with a warm and crafted writing style that anyone can understand and put to use. Distilled from other complex texts and real world experience, Michael Phillips delivers a gem when the planet needs it the most."

— **Tradd Cotter**, Mushroom Mountain; author of
Organic Mushroom Farming and Mycoremediation

"The world desperately needs the information in *Mycorrhizal Planet*! I am so glad Michael Phillips wrote this book. His approach is creative, inspired, and down-to-earth. A worthy effort with many useful practices laid out for all."

— **Dave Jacke**, coauthor of *Edible Forest Gardens*

"I firmly believe that the next big advancement in organic farming is learning how to harness the power of soil ecology by replacing mechanical tillage with biological tillage. *Mycorrhizal Planet* is an awesome book because it not only describes the importance of respecting living soil dynamics, it teaches how to act upon it. The chapter on practical nondisturbance techniques is especially enlightening to any serious market gardener."

— **Jean-Martin Fortier**, author of *The Market Gardener*

"Most land plants depend on symbiotic fungi in their roots — mycorrhizas — to help them to grow. Some, like orchids and many pines, depend on them absolutely. In fact, with no mycorrhizas: no land plants to speak of, and hence no land animals, including human beings. Like dung beetles and flies and microbes in general, root fungi are the largely unsung heroes of nature, cryptic creatures that make the world work. Michael Phillips's *Mycorrhizal Planet* brings them centre-stage — where, despite their modest demeanour, they deserve to be."

— **Colin Tudge**, founder, The College for Real Farming and Food Culture

"*Mycorrhizal Planet* is a thoroughly researched treatise on the impact of root fungi on the functioning of our biosphere. It is written in Michael Phillips's usual unique, enjoyable, and easily readable style. It is a must-read for all individuals seriously interested in the quality of human life and future of our planet."

— **George W. Bird**, professor, Michigan State University

"How lucky are we to be alive and growing plants right now? The humbling interconnectedness and relationships realized through *Mycorrhizal Planet* will fill you with wonder and have you questioning your role in the garden, orchard, or farm. This is the manual for upping your growing game!"

— **Eliza Greenman**, restoration orchardist and fruit explorer

"In *Mycorrhizal Planet*, Michael Phillips takes us on a journey into the realm of cutting-edge soil science, while always maintaining a playful sense of passion, excitement, and levity. As deep as Phillips goes into sharing his immense knowledge of the mechanics of vibrant living soils and their role in plant health, he never loses sight of the bigger picture — that of regenerating the planetary ecosystem. To that end, he offers robust practical applications for agricultural enterprises of all sizes. The beauty of this timely and important book is that we now know not only how but why we must embrace and cooperate with the innate intelligence of the biological world as we develop the agroecosystems that will sustain us in the future."

— **Scott Vlaun**, executive director, Center for an Ecology-Based Economy

"*Mycorrhizal Planet* awakens the reader to the interconnected, interdependent network of souls working on behalf of the earth right under our feet. The mycorrhizal fungi are our allies in promoting health for forests, orchards, and fields. Michael Phillips's comprehensive scientific knowledge, along with an abundance of practical information for the grower and a good dose of positive vibes for the future of our planet, make this new book one to add to your collection."

— **Linda Hoffman**, orchardist, Old Frog Pond Farm

Mycorrhizal Planet

How Symbiotic Fungi Work with
Roots to Support Plant Health
and Build Soil Fertility

MICHAEL PHILLIPS

Chelsea Green Publishing
White River Junction, Vermont

Project Manager: Patricia Stone
Project Editor: Benjamin Watson
Copy Editor: Deborah Heimann
Proofreader: Laura Jorstad
Indexer: Shana Milkie
Designer: Melissa Jacobson

Printed in the United States of America.
First printing February, 2017.
10 9 8 7 6 5   21 22

Our Commitment to Green Publishing

Chelsea Green sees publishing as a tool for cultural change and ecological stewardship. We strive to align our book manufacturing practices with our editorial mission and to reduce the impact of our business enterprise in the environment. We print our books and catalogs on chlorine-free recycled paper, using vegetable-based inks whenever possible. This book may cost slightly more because it was printed on paper from responsibly managed forests, and we hope you'll agree that it's worth it. *Mycorrhizal Planet* was printed on paper supplied by Versa Press that is certified by the Forest Stewardship Council.

Library of Congress Cataloging-in-Publication Data
Names: Phillips, Michael, 1957– author.
Title: Mycorrhizal planet : how symbiotic fungi work with roots to support plant health and build soil fertility /
   Michael Phillips.
Description: White River Junction, Vermont : Chelsea Green Publishing, 2017. |
   Includes bibliographical references and index.
Identifiers: LCCN 2016042187 | ISBN 9781603586580 (pbk.) | ISBN 9781603586597 (ebook)
Subjects: LCSH: Mycorrhizal fungi. | Symbiosis. | Roots (Botany)
Classification: LCC QK604.2.M92 P45 2017 | DDC 581.7/85 – dc23
LC record available at https://lccn.loc.gov/2016042187

Chelsea Green Publishing
85 North Main Street, Suite 120
White River Junction, VT 05001
(802) 295-6300
www.chelseagreen.com

Jerry Brunetti.
You walked the walk with cows and herbs
in the rolling hills of Pennsylvania.
Your green insights and cheerful tenacity
will long be appreciated.
Godspeed, brother.

CONTENTS

ACKNOWLEDGMENTS

Book ideas come at a guy and then it begins again. I wanted to learn more about fungal networks and healthy plants. Deeper understanding leads to better ways to garden, to orchard, to farm. And maybe, just maybe, sharing astonishing realizations about hyphal reach and robust phytochemistry will motivate more of us to become conscious of the importance of soil fungi. That seems to me the right starting point in working with living systems and ensuring a heritage for future generations. Such a journey requires having friends in the right places.

What began with incomprehensible musings on the writings of Rudolf Steiner ended with entirely good feedback from Peter Brady, Alan Surprenant, and Deb Soule.

Returning to home turf in Pennsylvania brought me to the Rodale Institute yet again. Congratulations are in order to Jeff Moyer, the hardworking farm manager rightfully made executive director. Research biologist Kris Nichols took me directly to the heart of soil aggregates and strengthened my fungal farming inclinations that much more.

Several mycorrhizal professionals helped me find important bits of the story line. Much is owed Mike Amaranthus at Mycorrhizal Applications, Graham Phillips at BioOrganics, and especially Jim Gouin at Fungi Perfecti. Steve Becker at Tainio Biologicals chimed in for the beneficial bacteria.

Posing questions to permaculture circles sparked insightful web conversations about mineral investment and ecological collapse. Checking in with Toby Hemenway confirmed my writer's voice was on track in getting points across clearly for those who care about such things as the humic mysteries.

Usha Rao at Ahimsa Alternative helped me better understand the critical role of fatty acids in plant health. Jason Hobson at Advancing Eco Agriculture sparked my interest in learning more about trace mineral cofactors. George Bird at Michigan State University provided important leads to better appreciate the role of foliar organisms in combating plant disease.

Apple grower friends have long been my go-to team for all sorts of pondering. Maury Wills, Harry Hoch, Linda Hoffman, Brian Caldwell, Pete Tischler, Eliza Greenman, Liz Griffith, and the veritable Mr. Bates all contributed an insight or two to this fungal ramble. The Holistic Orchard Network provided research funding for this project as well.

My cohorts in caring deeply for forest ecology go way back to earlier days writing for the *Northern Forest Forum*. Keep up the righteous work Jamie Sayen and Mitch Lansky. Meanwhile, down at ground level, I've had the occasion to make fungal friends in David Spahr, Tammy Sweet, James Nowak, and Greg Marley while out searching for edible mycorrhizal mushrooms.

Special thanks to Mary Anne Libby for permission to share a bit of Russell Libby's poetry. The organic movement lost a kindhearted visionary in his passing.

One of the fun things about writing a book is that you get to communicate with especially knowledgeable people. David Read has written about nearly every aspect of mycorrhizal symbiosis over a long, distinguished career at the University of Sheffield. This emeritus professor shared with me that his grad students have performed some especially important research. Cider was made from apples from trees inoculated with mutualistic fungi and from trees that were not. Tastings were done. Prodigious tastings, as a matter of fact. We can now irrevocably state that "arbuscular cider" holds glorious sway over the nonfungal swill. That alone may be all you really need to learn.

Next, a shout-out to the musicians who add a meaningful beat to this life. We start with a hot damn to Tom Paxton and those first "dual organisms" that crawled out of the slime. The final chapter stands as a tribute to the reggae master. Other lyrics are woven in along the way, even if subtly hidden in the context of kingdom Fungi. If you feel lost and lonely and don't know where to go . . . it can be kind of like finding Waldo. See? If you're singing along with James Taylor to "Lighthouse" . . . you are really too much in my head! Simply put, it was a way of keeping me in tune while writing earth melodies. Sometimes a guy just wants to have fun. That too! Honestly, you become like Fezzik, always coming up with a rhyme in *The Princess Bride*. You will have to venture down A1A to find 'em all. And indeed, that's your last clue.

No writer goes it alone. My team at Chelsea Green includes my original editor and dear friend, Ben Watson. Cheers, mate. Margo Baldwin keeps the sail ahead of the wind in finding inspiring writers to share sustainable insights. Pati Stone holds me to task when it comes to photos and illustrations and permissions. Elara Tanguy took my rough sketches and created fungal works of art. Frank Siteman continues to make it possible for me to stand in the best possible light. And still other folks helped make the design fall in place and the words ring true.

Those you live with bear the brunt of the writing journey . . . as there are reasons writers should be sent off on retreats, if only to protect the innocent. Gracie validated word choices and ran with my humor. Nancy brought in herbal support each day along with necessary perspective. Now I can gladly get back to renovating this old farmhouse and doing more together. My love to you both forever.

Lastly, thanks to all of you who keep asking the right sorts of questions. This discourse on the ways of symbiotic fungi and healthy plants has given me a chance to go that little bit further in explaining the biological side of holistic recommendations. The writing process leads to profound discoveries that make me a better grower as well. All gets easier as we go more fungal.

Fungal Consciousness

Mycorrhizal fungi have been waiting a long time for people to catch on.

Building soil structure and fertility that lasts for ages results only once we comprehend the nondisturbance principle. This alone guides the fate of an exceptional, yet too often forgotten, soil ally. Mycorrhizal fungi partner with the root systems of most every plant on our planet. The story behind this symbiotic relationship of fungi with roots launches all sorts of earth awareness. Carbon gets sequestered in more meaningful ways than "human offsets" will ever achieve. Healthy plant metabolism keys to balanced nutrient exchange delivered by fungal allies offering a savvy deal for sugars that plants create through photosynthesis. A profound intelligence exists in this underground economy that in turn determines the nutrient density of the foods we eat. Pick up a handful of old-growth forest soil and you are holding 26 miles of linear mycelium, most of it mycorrhizal.

Exploring the science of symbiotic fungi in layman's terms sets the stage for practical applications across the landscape. The real impetus behind gardening with mulches, digging with broadforks, shallow cultivation, forest-edge orcharding, no-till farming, low-impact forestry, and "everything permaculture" is to disturb the soil as little as possible. This in turn allows the fungal dynamic to thrive. Virtue lies in doing less so as not to screw things up, a tenet that I suspect most of us can handle.

Every biology book starts with scientific concepts. Just listen to people try to say *mycorrhizae* the first time when encountering this word, and it's obvious that a number of basics need to be covered. Those of us already applying fungal notions in working with plants know that the soil food web is important . . . yet we're not necessarily tuned into the full nuance of what's going on down below. Getting advanced about all this isn't so much PhD territory as it is appreciating how amazingly well plants and soil organisms "do" health. The complexity of enzyme synthesis and bidirectional protoplasm flow in mycorrhizal networking reflect Mother Nature in her finer hours.

We build on the work of many in tackling an elaborate subject like symbiotic fungi and plant health. New insights currently being gained through molecular exploration can be coupled with the dedicated observations of microbiologists, soil scientists, and plant physiologists over the decades. Reading through all the research has been quite a trip, and, well, let's just say you owe me one. I come at these vast fields of knowledge with a farmer's sense of the practical coupled with gratitude for science that offers a systems view. The parallels between human health and plant health just make this account all the richer. Thomas Jefferson has been described as having a limitless curiosity, backed by a deep faith in agrarian democracy. This nation's third president took into consideration a broad perspective in everything, from founding a university for higher learning to *the pursuit of happiness* as a natural right. We need more of this now. Those who know fungi must talk with those who know plants. Those who know phytochemical progression must talk with those who know the smell of rich earth in the furrow.

The facts and nothing but the facts come from the vast research laid out in *Mycorrhizal Symbiosis*, last updated by Sally Smith and David Read in 2008. The majority of unreferenced particulars presented in this work trace back to this primary source. More recent leads will be shared in endnotes, especially those providing important explanations of the issue at hand. Sorting through science papers is a language all its own (Phillips, 2017). Accordingly, every past history has origins (Phillips, 2017). Which of course makes everything a tad more dense (Phillips, 2017). But we do want to say thank you to each and all (Mom, 1965). And so we carry forward in this generation, triumphantly (Marley, 1980).

Our collective future pivots on many people coming to understand that soil fungi matter. That plant ecosystems must be respected. That soil

stewardship is our highest calling. Such has been said before, but now the jig is truly up. We either recognize the urgency involved at this critical juncture or the next century won't necessarily include the human race. Take these as fatalistic statements if you wish . . . or come on over to the other side.

The Nearings called it "the good life" and laid out a trail that Nancy and I picked up not long after college. Plant trees. Make compost. Sow cover crops. Harvest healing herbs. Build stone foundations. Chop wood. Let the chickens roam about. We share such pursuits with many of you. All sorts of amazing farmers, ranchers, gardeners, herbalists, orchardists, landscape professionals, and foresters are leading the charge. People are integrating livestock, market gardens, native plantings, and perennial crops into multifaceted systems. Goodbye monocultures, tillage, and bare dirt. Goodbye clear-cuts and monster skidder ruts. Rebuilding soil and increasing diversity go hand in hand with making a living doing authentic work that can be enjoyed. Equally important are the communities that support these efforts. From making local agriculture viable to organizing guided mushroom tours through community forests; from gardening up high on urban rooftops to remediating desperately hurting ground — we ourselves become healthier through every direct connection made with the soil that sustains us all.

Discovering the miraculous in the common would be the fungal slant on all this. The everyday ways of very small critters hold lessons for our species as far as planetary commitment goes. Honing one's ability to intuitively see the interaction between fungi and roots in the microscopic realm is how *fungal consciousness* takes hold. That we have things to learn from fungi isn't such a very big stretch to make. Hyphae reach out to nourish and communicate. Our neighborhoods would thrive if only people reflected the same sense of provision as found in any mycorrhizal network. Spores become the dreams we give our children. Life truly wants to go on.

Yes? Was that a question in the back? *Who am I to expound on the ways of mycorrhizal fungi and healthy plants*, you ask? No doctorate, no master's, no bachelor's degree in any such subjects. But for a found love of the soil—and more than a few fruit trees—I very well might never have ventured below. Days that began as a civil engineer on municipal construction projects inside the Beltway of this nation's capital shifted radically when I up and retired at age twenty-three to seek a more heartfelt path. Leafing through Thoreau, a rucksack of berries, and a classified ad in a homesteading magazine soon brought me to the White Mountains of New Hampshire. Long story short, I got my hands in the earth.

You become a "plant person" by discovering a passion within yourself for growing, well, just about everything. A "tree person" happens to go a bit further out along the limb, that's all. Enter the whims of microbe eating microbe, that very first dipping of bare roots into mycorrhizal inoculum, and the dawning realization that orchard trees want to grow in a fungal duff ecosystem. Root tips proceed behind a vanguard delving into soil aggregates and rock crevices where little other than the finest hyphae can go. Ever notice how the universe only grows bigger as you zoom in on minutiae? I was hooked on learning more about the soil food web and what makes for long-lasting soil fertility. Back on the receiving end, plants prove healthier when you honor that Nature knows best. Put another way, those apple trees of mine proved to be the very best teachers. Nancy's work with medicinal plants opened my eyes to the marvel of healing synthesis, only for me in the context of what a guy needs to get by. Plants and fungi become that much more resilient by looking out for the other. The resulting phytochemical cascade in turn serves both the farmer and the herbalist. Worlds merged yet again as fascination led the way.

Are you getting a feel for how this works? Thinking fungally, exploring fungally, intuiting fungally. We're each quite capable of taking what's been gifted to us and going further still.

Trusting that journey makes it possible for me to share practical teachings about mycorrhizal fungi and real-time plant health. The two are intimately linked in ways that make a holistic approach all the more spot-on. I always liked what Liberty Hyde Bailey of Cornell had to say, slightly paraphrased: *If a grower knows why, he or she will teach themselves how.* You're about to get a good dose of plant wisdom interspersed with some guy humor to keep things hopping. Jump around; take what's useful; have fun. Every growing season is a renewed opportunity to let the green world astound.

We start off with a requisite shot of science to launch a fuller understanding of mycorrhizal fungi and healthy plant metabolism. Chapter 1 introduces the relevant fungal concepts upon which practical application will follow. Knowing about species adaptability, propagules, and nutrient transfer mechanisms makes us better earth stewards, frankly. The lure of the "fungus-root" comes into its own once we appreciate it's not about green plants alone. Chapter 2 brings in aspects of plant physiology and organic chemistry to get at the heart of phytochemical progression. Sounds daunting, perhaps, but know this has been finely tuned to help you better understand the natural world. All starts to click as growers recognize why investing in health is far preferable to dealing with the myriad of problems that arise as a result of mistreating soil life. I *double dog dare ya* to embrace holistic means in growing food for your family and community . . . as of course you will after taking in this wholesome plant conversation. Working with resistance metabolites and fungal allies beats toxic thinking hands down.

That front-end dive into science allows us to further explore symbiosis benefits as we move along to consider the wider ecosystem. Chapter 3 explores how entire plant communities are linked by far-reaching mycorrhizal networks in an underground economy. The language of trade helps in understanding how deals get done. Profound intelligence exists in natural systems, so prepare to be awed. And then prepare to not disturb the impeccable! Chapter 4 shares ways that we as growers can specifically favor the fungal dynamic and thereby create good soil tilth. Soil aggregates are the direct result of respecting the primary role of fungi in what might best be thought of as biologically induced alchemy. We'll get down and dirty in chapter 5 when it comes to inoculum nuance and traditional techniques to further the fungal cause. Pay particular attention to the "carbon pathways" that in truth are our one saving grace in these tenuous times.

Weaving an integrated tapestry of health on land we care about comes through actual doing. Chapter 6 finds us outside and raring to get to work. Thinking fungally puts the right framework on how to garden, design landscapes, and grow tree fruit. Fostering fungal precepts makes for more resilient forests and farms. Those who love nuance as much as I do will enjoy the fungal riffs to be underscored in every deed.

Inspiration for the fungal gourmet will be found in chapter 7. Finding the edible among mycorrhizal mushrooms in the wild is a passion unto itself. Lastly, all comes round in chapter 8 with a big-picture view of what a world versed in carbon flow might look like. Our species has some daunting choices to make, but in the end it will come down to each one of us finding joy in fungal ways. Transformation happens when we take the initiative to do better by soils everywhere. The fungi will take it from there.

This notion that "plant toes being tickled by fungi" bestows life to our dear planet—and so many two-leggeds don't even have a clue—means the time has come to tell the full story about mycorrhizal fungi.

CHAPTER 1

Mycorrhizal Ascendancy

The first plants to make the move from the sea to land did so without roots, relying instead on a dynamic relationship with fungi to bring nutrients and water for growth. This dual-organism approach proved to be a pretty good idea, evolutionarily speaking. Some 450 million years later, most plants continue to dance with fungal partners.

Symbioses are intimate associations between two distinct organisms. The mutualistic relationship between plant roots and fungi goes largely unseen, yet pulses through the top inches of healthy soil everywhere. This collaboration is one of several that make possible life on earth. Nor should we be surprised. The human perception that making it in this big ol' world depends upon putting the individual first is common enough. *That phosphorus is totally mine, by gar*. Yet plants and fungi thrive because both have entered into an enduring arrangement whereby certain plant essentials are provided in exchange for certain fungal essentials.

Combining the Greek words *myco* for 'fungus' and *rhiza* for 'root' provides the name for this pairing. The term *mycorrhiza* refers to the coming together of one fungus with the root system of one plant. Let that combination of *fungus-root* take hold in your imagination. A linguistic fork in the road comes up immediately thereafter: The customary way to make this into a plural in North America has been the latinized *mycorrhizae* (pronounced my-co-RIZ-ee), whereas the British pluralize the word by simply adding an *s* in referring to *mycorrhizas*. Both work in a congenial world. These terms always denote the conjoined structure. More confusion enters in when we discuss the individual partners involved. A *mycorrhizal fungus* joins with a *mycorrhizal host plant* to form a mycorrhiza. Subtleties abound. Yet you may well catch me out on occasion . . . for now let's get back to that coming from the sea bit.

Crawled out of the slime is more like it. The algaelike ancestors of plants dwelling in the primordial sea experienced quite a shock when waves lapped emerging shorelines. Land ho! The foremost possibility was to shrivel up and die. Hanging out in a tidal pool was okay until things dried out. Now and then it chanced that "someone else" shared the tidal pool. The earliest fungi have oceanic origins as well, as seen in a modest fossil record.[1] The precursors of vascular plants lacked roots but had the ability to manufacture (photosynthesize) simple sugars; the fungi possessed hairlike filaments capable of absorption but needed the energy associated with carbon chemical processes. Negotiations began . . . and soon enough, as these things go across the millennia . . . a surrogate root deal materialized. Former tidal pools were kept replenished with rainwater. Lignins were synthesized that could support upright structure. Mineral particles accumulated around emerging root nubs (rhizoids). The fungus served as both anchor and nutrient supply chain throughout these shifts. No one moved, and yet everything seemed to move as the sea receded further.[2]

The partnership proved resilient. More to the point, the extra energy generated for these first

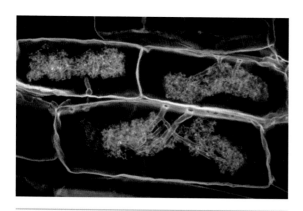

Here we see arbuscular mycorrhizal fungi penetrating into the cells of a root. Look closer. Perhaps you see miniature trees. Flip that perspective and you see the feeder root system of most any plant. Then again, your view may be more anatomy-oriented, as these fungi also resemble the alveoli in our own lungs. Herbalists speak of the "doctrine of signatures" underlying a plant's healing gifts. Similarly, mycorrhizae make clear an essential connection to life itself. Courtesy of Larry Petersen, University of Guelph.

plants (in being supplied critical nutrients) allowed for greater differentiation and development of complex tissues. Leaves were next, followed by flowers and seeds. Tall trees came to be. Vines. Edelweiss. Healing herbs. My apple orchard. All releasing oxygen, all enabling more life.

We indeed have much to celebrate in pondering this fungal connection between healthy soil and healthy plants. Mycorrhizae present a system of synergy in that both partners gain more than would be available to the individual. Little wonder that symbiosis is the rule rather than the exception across the globe. All told, about 92 percent of plant species studied (of approximately 80 percent of the total number of plant species) share an affiliation with different types of mycorrhizal fungi. And those that don't—the so-called nonmycorrhizal plants—very likely lost it through some combination of extreme habitat and parallel evolution.

Mycorrhizal fungi form a filamentous body called a *mycelium*. This is composed of branching threads called *hyphae* that grow through and across substrates to access food resources, secreting enzymes that catalyze chemical reactions and break down complex matter into simple compounds, which can then be absorbed back through the cell wall. The reach of the mycelium of a single fungus beyond the root system effectively enlarges the surface-absorbing area of the plant partner by ten to one hundred times. The uptake of water and nutrients (such as phosphorus and zinc) by the fungus is delivered to the root by means of an internal mycelium in exchange for simple sugars and synthesized compounds from the plant. This fuels the growth of both partners. Certain mycorrhizal fungi can be seen on tree roots, yet more often than not, symbiotic species on the majority of vascular plants will be invisible to the naked eye.

The fineness of hyphae in comparison with the relatively blunt hairs on feeder roots reveals how mycorrhizal fungi can access diverse nutrient niches. We'll literally be peering into minute soil aggregates and carbonized biochar particles and even solid rock in considering the hyphal perspective. Let's zoom out first to a macro view of this scene. Beneath the imprint of one's foot, extending down into the soil but mostly meandering nearer the surface, are 300 miles of fungal hyphae.[3] Inconceivable. I could walk from our farm in the White Mountains of New Hampshire to Montreal and back along the gossamer filaments found under a single footstep.

Now let's zoom up another level and take a planetary perspective. Beneficial patterns show up consistently throughout creation. A plant cell membrane surrounds the interior workings of each cell, be it in the root, the stem, or the leaf. Similarly, our very selves depend on a community-wide symbiosis with some hundred trillion microbes that form a protective membrane both on the surface of our bodies and on every square centimeter of our respiratory and digestive tracts. The vast network of mycelia lying just beneath the surface of the soil (and mind-blowingly enough, throughout the sediments on the ocean floor) is the planetary membrane holding life's sacred trust. Mycorrhizal fungi may seem to be a "small matter," but together with the full coterie of soil biology and plants they are the wellsprings of life as we know it.

The majority of biological activity takes place within the litter layer and the top layer of actual soil below it. The organic duff zone is known to soil scientists as the O horizon and the rich soil immediately below as the A horizon. Typically the latter ranges from 4 to 16 inches (10–40 cm) deep. Here can be found the highest densities of plant roots and soil organisms when compared with anywhere else in the soil profile. We will be delving into this web of life where microbe consuming microbe becomes just as important as the friendly exchange between mycorrhizal fungi and plant roots.

Not all nutrients for plants are garnered by means of fungal enzymes and subsequent hyphal trade. Fungi fall prey as well in the great cycle of assimilation and mineralization — though nowhere near as readily as bacterial cohorts. Polysaccharides compose the fungal cell wall, and chief among these compounds is *chitin*. This durable polymer adds rigidity and structural support to thin hyphae, thereby giving fungi more staying power in nondisturbed ground. Protoplasm flowing through these essentially open pipelines delivers proteins and other essential reserves to further tip formation into new territory. Mycelium grow rapidly, clocking an average hyphal pace of just over 1 micrometer per hour on a good day. (Everything is relative!) This outreach ability of mycorrhizal hyphae makes possible extension into untapped spaces.

Mycorrhizae are the normal nutrient-absorbing organs of the majority of plant species. *Normal*, as in nobody does it better. Plants and fungi are equal partners in this venture. Each species of mycorrhizal fungi offers slight variations on the plan, while plant specificity tends to be wide open. Consequently, interactions between diverse populations of mycorrhizal fungi and plants shape the structure and function of ecosystems as a whole.

A listing of symbiosis benefits to plants reads like an incredible marketing pitch from start to finish. Mycorrhizae improve plant health. Some plants receive no net nutrient benefit but gain pest resistance instead. Mycorrhizae enhance a plant's ability to tolerate environmental stress, be it drought or high soil temperatures or metal toxicity. Mycorrhizae help make food crops nutrient-dense. Throw in soil stabilization and ecosystem resilience for good measure. Oh yeah, there's one more

SYMBIOSIS BENEFITS TO PLANTS

NUTRITION

- Increases surface area of nutrient uptake
- Unlocks phosphorus for plants
- Acquires nitrogen from organic matter
- Improves uptake of trace minerals
- Enhances nutrient density of crops

SEEDLINGS

- Prevents "damping off" disease
- Reduces transplant shock
- Supports root initiation on cuttings

FIELD AND FOREST

- Stabilizes soil aggregates
- Sequesters carbon
- Improves plant growth and yield
- Delivers moisture as needed
- Augments deeper root penetration
- Suppresses root pathogens

PRACTICAL ADVANTAGES

- Mediates heavy metal toxicity
- Helps plants cope with soil salinity
- Breaks up subsoil compaction layer
- Suppresses nonmycorrhizal weeds
- Cuts fertilizer requirements
- Improves tolerance of higher soil temperatures

HEALTHY PLANT METABOLISM

- Improves rate of photosynthesis
- Plays cofactor role in protein synthesis
- Provides reserve energy in lipid form
- Stimulates induced systemic resistance

MYCORRHIZAL NETWORKING

- Ensures balanced nutrient uptake
- Ensures healthy forest succession
- Facilitates plant-to-plant communication
- Provides the foundation for ecosystem resiliency

little thing. Root fungi provide *the solution* for our current climate change impasse. Free of charge. No carbon credits involved. All we humans have to do is get our heads out of our muck and stop destroying natural systems under the guise of unmitigated greed. Mycorrhizae sequester carbon. Big time. Here lies a future and a hope.

What I'm designating as "mycorrhizal ascendancy" isn't about fungi making a bold move. Nor do plants need to do anything more. It's you and I who must become reasonable tenants on this one holy earth. Attention, *Homo sapiens*. It's time to explore the vastness of the mycelium. What follows sets the stage for practical application, should you be the type of person who likes knowing the science that informs fungal stewardship.

Mycorrhizal Types

Symbiotic fungi are identified by anatomical form and the nature of the association with plant roots. The first thing to be learned in distinguishing these

The Pace of Mycorrhizal Discovery

Humans took note of the association between fungi and the roots of plants in the middle of the nineteenth century. A Polish botanist named Franciszek Kamieński recognized this union as a symbiosis in 1881. Further research was carried out by Albert Bernhard Frank, who introduced the term *mycorrhiza* in 1885, while investigating truffle production for the king of Prussia. Improvements made around this time to the compound microscope allowed observers to see hyphal interaction with roots on a cellular level. *Revue Generale de Botanique* published a series of outstanding drawings detailing endomycorrhizae made by I. Gallaud in 1905. Decades would pass as conventional thought slowly adapted to the idea that fungi could be something other than bad for plants.

Functional discoveries began to come along by the 1950s. Early research focused principally on nutrient uptake and the formation of ectomycorrhizae. Attention began to shift equally toward arbuscular mycorrhizae by the 1970s. Published papers started coming at a consistent pace as soil scientists and mycologists built on a foundation of knowledge. Field studies began to encompass an ecological point of view by the 1990s. More recently, insights gained through molecular biology have opened windows into the taxonomic relationships of species and (shamefully) genetic manipulation. Compartmentalized root chambers have revealed more about the complexities of nutrient transfer and mycorrhizal networking.

Great dedication has gone into figuring a few things out about the robust web of life beneath our feet. What's wonderful is realizing there's so much more yet to be discovered about the fungal underpinnings of plant health.

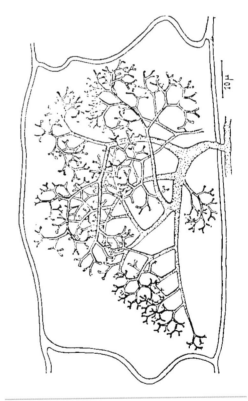

The microscope provided a first view of arbuscular fungal-root interactions a mere century ago. Illustrations by I. Gallaud (1905).

fungi by type is that there's a whole lot of shaking going on. Tree-oriented ectomycorrhizal fungi sit on one end of the spectrum with arbuscular mycorrhizal fungi on the other. The boundaries in between are less discrete as fungal rubrics are put to the test. Full genome sequencing has broken the classification scene wide open all the more as researchers correlate genetic ties among species.

Hyphal structure of these fungi consists of two basic parts. Configurations of filaments that develop separately from the physical root space have a different mission than those that work in concert with the root interface. The *extraradical mycelium* explores for nutrients out beyond the root to carry back to the plant, whereas the hyphal structure alongside or within the root needs further clarification. Some fungi completely cover root tips with interwoven hyphae and restrict cortex penetration to between cells. Others develop a full-blown *intraradical mycelium* and poke directly inside root cells for nutrient exchange. The prefixes *ecto* and *endo* take into account whether the action is 'without' or 'within' the root cell. Hook those up and voilà! The terms *ectomycorrhizae* and *endomycorrhizae* provide the

broad beginnings by which to consider this corner of the kingdom Fungi.

The five types grouped under "endomycorrhizal fungi" stretch these definitions when mechanisms and plant hosts are distinguished yet again. Very dissimilar associations fall under this banner. Some of these fungi specialize in orchids, others in heath-loving plants. The largest functional group by far is the arbuscular type of mycorrhiza. This last group is so dominant in the plant kingdom that people often simplify the whole discussion by giving these fungi primary rights to the *endo* prefix.

Arbuscular mycorrhizae (AM) associate with the roots of four out of every five plant species, including many crop plants. AM fungi may well have the ability to invade underground parts of all terrestrial plants; they are the fungi that assisted plants in moving to land in the first place. Jim Gerdemann put it this way back in 1968, "The symbiosis is so ubiquitous that it is easier to list the plant families in which it is not known to occur than to compile a list of families in which it has been found."[4]

Intercellular hyphae of AM fungi first develop trunk lines between the outer cells of root tissue.

Table 1.1. Mycorrhizal Types

Endomycorrhizal Fungi (within)					Ectomycorrhizal Fungi (without)	
ARBUSCULAR	**ERICOID**	**ARBUTOID**	**MONOTROPOID**	**ORCHID**	**ECTO-**	**ECTENDO-**
Aseptate	Septate	Septate	Septate	Septate	**Septate**	Septate
Intracellular structure	Yes	Yes	Yes	Yes	No	Yes
No	No	Sometimes	Yes	No	**Mantle**	Sometimes
Vesicles 80% of species	No	No	No	No	No	No
No	No	Yes	Yes	No	**Hartig net**	Yes

Source: Adapted from David Moore's World of Fungi website.

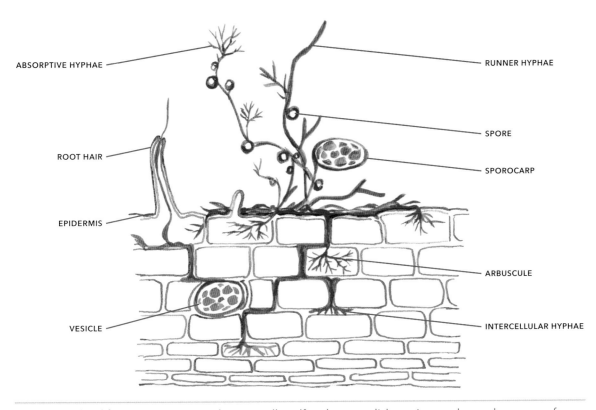

Endomycorrhizal fungi penetrate into the root cell itself and accomplish nutrient exchange by means of arbuscule structures and intercellular hyphae.

These veer off to penetrate cortical cells and then form microscopic *arbuscules*, whose fine branching form resembles little trees (hence the name). Nutrients are exchanged here between fungus and root. Most AM fungi go on to form storage structures called *vesicles* within the colonized root as well. These contain large amounts of lipids, and often numerous nuclei. Vesicles left behind in root fragments are one of the means by which these fungi carry forward to infect new roots. We'll be learning more about ongoing colonization by means of an assortment of fungal strategies just ahead.

There are very few cross walls (septa) in AM fungal hyphae. This "aseptate nature" evidences the primitive origins of this grouping, as no other mycorrhizal fungi share this trait. Nonseptate hyphae are formed by very long cells containing many nuclei. Cross walls form at branch points that connect one filament to another, preventing the entire network from being compromised if one hypha is injured. Protoplasm flow is bidirectional and unhindered as nutrients and nuclei alike readily pass through pores at each branch juncture.

Arbuscular-type hyphae outside the root are just as invisible to the naked eye. Those filaments that first exit the root are known as *runner hyphae*, which in turn develop thinner, branched networks of hyphae to absorb nutrients from the surrounding soil to send back to the root. This outboard reach of an individual AM mycelium encompasses up to a hundred times more soil volume than

Development of extensive mycelium and rhizomorphs by *Suillus bovinus* in symbiosis with a seedling of *Pinus sylvestris*. Note that the germinants to either side have already plugged into this supporting network. Courtesy of David Read, University of Sheffield.

feeder roots alone . . . thereby providing access to nutrients from as much as 3 inches (7–8 cm) away from the host roots.

Ericoid mycorrhizae grow in habitats where the majority of nutrients are in organic form only. This is a "nutrient niche" not available to nonmycorrhizal plants or arbuscular mycorrhizal plants. Plants belonging to the order Ericales include acid-loving berries, heather, and evergreen mountain laurel.

Loose hyphal networks form around the outside of hair roots, which then penetrate root cortical cells to form *intracellular coils* for nutrient exchange. Extension of ericoid hyphae into surrounding soil is relatively limited, thus making availability of moisture all the more critical for these sorts of plants.

Two other identified groups of fungi tag along with these same plant groups, only they take on some ecto-type features as well. **Arbutoid mycorrhizae** tend to form a sheath of hyphae around root tips and use intercellular structures as well as coils to exchange nutrients. Uva-ursi and related shrubs like alpine bearberry have this arbutoid affiliation. **Monotropoid mycorrhizae** tap into hyphal systems of tree fungi to assist species like Indian pipe (*Monotropa uniflora*) in a similar coiled fashion. Such heterotrophic plants essentially transform unsuspecting neighbors to redirect sugars from fungal purpose for "ghost plant growth" in lieu of being able to photosynthesize directly.

Orchid mycorrhizae help plants in the family Orchidaceae pass through a germination and early development phase when they are completely dependent on an external supply of nutrients and organic carbon. Many orchids lack chlorophyll until they are fully established and thus must rely on fungi to get started. These species of fungi verge on being parasitic by simultaneously colonizing photosynthesizing plants to effectively deliver the goods to the fledgling orchid. Point of view makes all the difference.

Ectomycorrhizae (EM) associate with dominant forest trees of many types, amounting to as much as 3 percent of all terrestrial plants. The relative count of host species may be small, but keep in mind that trees cover a disproportionate amount of the earth's surface overall. Accordingly, ectomycorrhizal networks can be quite vast, accounting for a third of total microbial biomass on forest soils.[5] Some ecto-oriented plants are not exclusively colonized by EM fungi, which opens the door to even more fungal interaction.

The ability to secrete hydrolytic enzymes that degrade organic molecules to secure nitrogen reveals the dual nature of EM fungi. Saprotrophic (decomposition) fungi are not mycorrhizal as a rule. On the other hand, mutualistic fungi must be able to enter into root space to share these nutrients without triggering plant defenses. Outright saprotrophs are unable to do this. The explanation lies in the distant past. EM fungi and trees share similarities across all continents, suggesting that this symbiosis began some 160 million years ago.[6] Numerous gene lines in EM fungi trace back to white and brown rots (see page 72 in "Soluble Lignins") noted for serious breakdown jive.[7] Apparently, some of the early decomposition fungi to come on the tree scene (as organic matter began

to accumulate) switched modes and evolved a symbiotic behavior. These fungi lost the ability to degrade plant cell wall polysaccharides (principally cellulose and pectates) while gaining the ability to enter intercellular spaces in the root in exchange for mineralized nutrients.

This developmental pathway brought about EM fungi with entirely different traits from those of AM fungi. The *mantle* (fungal sheath) of interwoven mycelium on the surface of the finest roots is often visible to the unaided eye or by use of a hand lens. The fungi fit to the root tips like fingers on a glove. An internal network of hyphae, known as the *Hartig net*, weaves between the cells in the root. The penetration depth of this nutrient exchange structure differs between angiosperms and gymnosperms. Most angiosperms (deciduous species) develop a Hartig net confined to the outer epidermis, which is often radially elongated. By

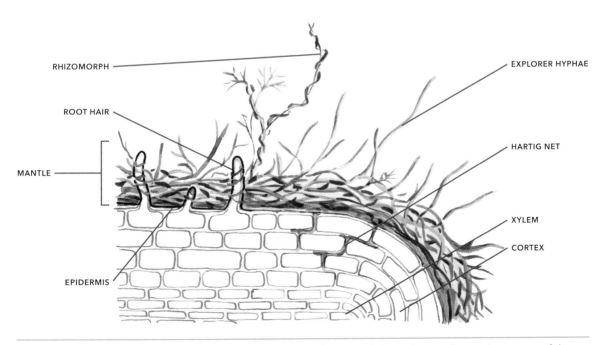

Ectomycorrhizal fungi sheath the outside of tree roots and accomplish nutrient exchange by means of the Hartig net structure reaching between root cells.

contrast, the Hartig net in gymnosperms (coniferous species) normally penetrates further within the cortical cells and may even extend to the perimeter of the root core.

Fine root tips are typically found in topsoil layers containing humus — rather than in underlying layers of mineral soil — and this holds true for the bulk of an ecto-type mycelium. This correlates well with the layers of decomposing litter on the forest floor. Mycorrhizal short roots grow outboard in search of nutrients by elongation and branching. Conifer roots with ectomycorrhizae have subdividing branching patterns, while deciduous trees tend toward parallel branching. Singular strands or masses of long-distance hyphae (called *rhizomorphs* when tightly bound together) also form, providing these fungi with far more nutrient reach than arbuscular species. Such *explorer hyphae* are most often found where root density dissipates on the forest's edge, reaching out as much as 12 feet (4 m) beyond tree-rooted origins. Others drive downward in search of untapped mineral resources in rock.

Many of the EM fungi are mushroom-forming species, including highly prized edibles such as chanterelles, boletes, and matsutake. Stay tuned for an overview of symbiotic Basidiomycota that can be enjoyed at the dining table in chapter 7. And should you find truffles, those Ascomycota that form underground fruiting bodies in association with certain tree species, consider yourself to be in the money.[8] Healthy fungal forests are about far more than marketable board feet of timber.

The "last fungus standing" is more a catchall for species exhibiting adaptability on all fronts. **Ectendomycorrhizae** have been given a purely descriptive name for root interactions that exhibit characteristics of both ectomycorrhizae and endomycorrhizae. These are essentially restricted to the plant genera *Pinus* (pine), *Picea* (spruce), and, to a lesser extent, *Larix* (larch). Direct penetration of living cells in the host root with otherwise ecto-type characteristics is quite the lead-in for where we're going next.

Fungal Adaptability

One recurring theme right out of mycorrhizal central casting suggests that we can always expect broad innovation from the fungal kingdom. A wide diversity of structure, development, and function exists in any single mycorrhizal type. A species that thrives in boreal regions can just as likely be found in the tropics. Isolated strains may behave somewhat differently, perhaps testing bounds with a new plant host or adjusting enzyme synthesis to free a trace nutrient from soil. This tremendous ability to adapt to numerous environments has allowed symbiotic fungi to thrive.

Soil microbiologists have long recognized some 150 species of arbuscular mycorrhizal fungi that colonize the root systems of plants, and this is after decades of careful classification. That long-standing count has recently been shown to account for only half of what's out there, thanks to an exhaustive sampling of roots from all continents (excepting Antarctica) by means of molecular analysis.[9] But still. Some 300-plus AM symbionts are known to affiliate with over 250,000 species of plants. Functional aptitude clearly lies on the side of the fungi. Somehow a relatively small number of organisms have managed to keep the upper hand in untold settings for a very long time.

Arbuscular mycorrhizal fungi are asexual and thus clonal . . . yet how does one family avoid large numbers of deleterious mutations over hundreds of millions of years? Pressures for change are small in a mutualistic symbiosis compared with a parasitic interaction where host and foe are continually trying to outmaneuver the other. Consistency across generations suits AM fungi precisely because the acquisition of carbon from the roots of plants in exchange for nutrients likewise suits its trading partner. Nature simply nailed it from the beginning. No need to rotate out players when each *fun guy* (so glad to get *that* out of the way) already has the skill set needed to deal with climatic curveballs.

The ability to adapt by improving on individual performance without taking on an entirely new genetic identity involves teamwork. Fungal hyphal systems are capable of fusing together. When the tip of one arbuscular hypha comes into contact with the hypha of a genetic twin, regardless of where along its length, the two organisms can merge together. This joint undertaking is known as *anastomosis*. Such plumbing alterations can be done "in house" as well when hyphae of the very same fungus merge together to keep continuity intact.[10] Protoplasmic flow can then be routed in multiple directions, to multiple plants, even transferred within plants to other hyphal systems of entirely different origin. All without restriction. What's known as the *common mycorrhizal network* can literally link the roots of all plants at a site. This is deserving of attention in its own right (see "Mycorrhizal Networks," page 55). Schmoozing notions carry even further with certain tree species that "bridge" the gap between different fungal types (see "Bridging the AM-EM Divide," page 152) in the name of orchard health. For now, let's simply focus on what it means to be able to share resources by spontaneous anastomosis.

Dissolved nutrients, recognition hormones, and nuclei can be passed from one hypha to the next. Dig that last bit. Any and all adaptations by a fungal organism will be recorded in its genes. Fungi pack vast numbers of nuclei into both hyphae and spores alike, ranging from eight hundred to up to thirty-five thousand at a pop. An isolated strain playing the numbers can advance "subtlety of function" without outright altering species identity. Scientists have only recently recognized that anastomoses can form between genetically distinct isolates of the same AM species, albeit at a low frequency.[11] Nuclei mingle in common protoplasm, passing along improved traits, building an arbuscular portfolio for what we might well call the Fortune 300.

Sexual mingling of ectomycorrhizal hyphae results in fruiting bodies that do cast forth potential new types. Recognized species of EM fungi number around twenty thousand, yet these affiliate with only a third as many plant hosts. Convergent origins with decomposers over a long evolutionary period assure a broad gene pool in ectomycorrhizal species. Trees that have been introduced from one place to another (such as eucalyptus, which was brought to North America from Australia) serve as hosts for invasive fungal species that potentially can outpace native species.[12] Sometimes the aggression of the so-called invasive plant is foremost about soil negotiations below.

That underlying metaphor applies when we consider the ecological adaptation of mycorrhizal fungi across the landscape. Such "fungal tracking" takes into account soil qualities and the types of plants likely to be found in particular places . . . and thus the types of fungi in association with those plants. Ericoid mycorrhizae are predominant in highly organic soils in the far north, where plant growth is nitrogen-limited. Walk across alpine heath or straddle a bog, see plants in the Ericaceae family (such as blueberries) — and *you know who* will be down in the root zone. Ectomycorrhizal

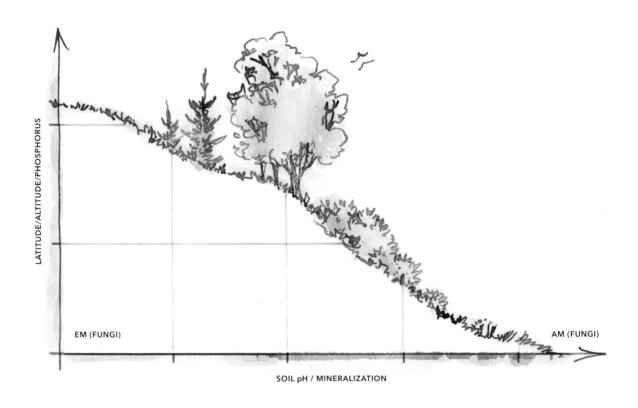

The influence of soil type, phosphorus availability, mineralization, and latitude provides an ecological perspective to mycorrhizal colonization. Adapted from D. Read and J. Perez-Moreno, "Mycorrhizas and Nutrient Cycling in Ecosystems," *New Phytologist* 157 (2003): 475–492.

fungi reign in nitrogen-limited forest ecosystems with appreciable litter accumulation on the surface and relatively few plant species. Hike through the woods, especially at intermediate altitudes throughout the middle latitudes, and most every coniferous and deciduous tree will be bopping to ecto vibes.

Arbuscular mycorrhizae dominate in phosphorus-limited locations. Transitions are made as the forest opens up to herbaceous and woodier undergrowth. Species-rich ecosystems are always far more arbuscular. This same fungal imperative carries out onto the savanna and across the prairie into shrubland and grassland ecosystems.

Tropical ecosystems featuring a diverse range of plants have strong arbuscular leanings as well. You can walk some distance through a dripping rain forest before encountering two individual trees of the same species. Compare this with where we started . . . the species mix in the boreal forest may be limited to four or five tree species at most. Yet these aren't the only fungal clues. Look around for mushrooms. Mycorrhizal fruiting bodies aboveground come about solely through ecto-type

Table 1.2. Plant Affiliations

Mycorrhizal Type	Major Groups of Plants	Plant Species w/Affiliation	Fungal Phylum	Estimated Number of Fungal Species
Arbuscular mycorrhiza	Most herbs Most vegetables Grasses Legumes Cane fruit and ribes Fruit trees Many forest species Horn and liverworts	250,000	Glomeromycota	>300
Ectomycorrhiza	Pine, spruce, fir Oak Willow, poplar, birch Eucalyptus Beech, maple Some Rosaceae Dipterocarpaceae	6,000	Basidiomycota and Ascomycota	20,000
Orchid mycorrhiza	Orchids	35,000	Basidiomycota	25,000
Ericoid mycorrhiza	Blueberry Cranberry Heath Huckleberry Lingonberries Rhododendron	3,900	Mainly Ascomycota, some Basidiomycota	<150

Source: Adapted from M. van der Heijden et al., "Mycorrhizal Ecology and Evolution: The Past, the Present, and the Future," *New Phytologist* 205 (2015): 1407.

affiliations. Few trees in the tropical forest form ectomycorrhizae, while in the northern forest all trees form ectomycorrhizae. As a result, the northern forest ecosystem features a great variety of large mushrooms, while the tropical one supports fewer species of mostly small mushrooms.

The elephant in the room when it comes to fungal adaptability is plant specificity. Fungi are broad generalists with regard to choice of partners . . . but what do plants "think" about all this?

Plants have varying degrees of dependence on mycorrhizal associations. This can be assessed by examining the proportion of the root system involved with symbiotic fungi. Researchers will estimate *mycorrhizal root length* in terms of percentage of colonization of the plant's root system, along with a who's who analysis of the fungi involved. Such attempts at exactitude will vary at different points in the season yet still reveal consistent differences in the intensity of mycorrhiza formation from plant to plant. Species that generally have high levels of colonization are noted to be obligatorily mycorrhizal. Those with variable levels are considered facultatively mycorrhizal.

NONMYCORRHIZAL PLANTS

Nonmycorrhizal plants resist colonization and are able to function effectively without the association. A number of these fall into the category of *ruderals*, the early succession plants that reclaim disturbed ground. Others are familiar vegetables and particular cover crops. Many parasitic and carnivorous plants, along with rushes and sedges that develop cluster or dauciform roots, are included here as well. See www.mycorrhizas.info/nmplants.html for a full listing of nonmycorrhizal plants.

AMARANTHACEAE

- amaranth
- beets
- chard
- goosefoot (lamb's-quarters)
- redroot pigweed
- spinach

BRASSICACEAE

- broccoli
- cabbage
- cauliflower
- kale and collard greens
- mustard
- rutabaga and turnip
- tillage radish

MONTIACEAE

- claytonia

POLYGONACEAE

- buckwheat
- dock
- knotweed
- rhubarb
- sorrel

PORTULACACEAE

- purslane

SCROPHULARIACEAE

- figwort
- mullein

In natural ecosystems, plants with "on-again, off-again" associations are more common in very dry, wet, or cold habitats. Overall productivity tends to be constrained in these extreme situations. Nonmycorrhizal plants do best in disturbed habitats. This surely tells us something about "weeds and what they tell" regarding soil biology.

A single root system often has multiple affiliations with different mycorrhizal fungi. Conversely, different plant species share choice of fungal partners. Those are the founding principles behind mycorrhizal networking based on distributing surplus resources community-wide. Think a minute about the incentive implied in that statement. Any plant inclined toward "one fungus to rule them all . . . and in the darkness bind them" will be at a disadvantage with plants open to broader societal benefits. Yes, a plant may be better off associating with one symbiont for a time rather than none at all. That singular fungus, meanwhile, isn't necessarily keen on keeping the credit all for itself. Conditions change, species shift, and eventually the dominant paradigm will be restored. There's no such thing as "trickster fungi" in a healthy soil ecosystem that will stop at nothing to hold a niche.[13] Nor do plants have any reason to support an insurrection.

Ask Scots pine (*Pinus sylvestris*) about its 117 species of known mycorrhizal associates and you'll come to understand there's a party going on.

Propagules

Fungi carry forward to the next generation by means of spores, hyphal fragments in roots, and actively growing mycelia. The nuance of propagules will come up repeatedly as we consider techniques to abet fungal diversity. These are the structures that produce a new fungal organism to partner with yet more plants.

The most active kind of propagule is the living mycelium itself. Most new seedlings in healthy native vegetation are colonized in this way, as are plants on the periphery of an expanding fungal reach. Hyphae never pass by an opportunity to establish new friendships.

The pace of runner hyphae is nothing less than insistent. The mycelia of *Glomus mosseae* spreads outboard at a modest 3 micrometers a day in a soybean field. Other species average closer to 20–30 μm of reach from one sunrise to the next. The longest recorded growth spurt has been 90 μm in a single day when soil conditions prove just right. Hyphae pass through narrow soil pathways, practically sprinting along a wide-open worm tunnel.

Fragments of roots usually contain live fungi, allowing extant root pieces to be very effective propagules as well. Certain arbuscular-type fungi sporulate within the root cortex, thereby embedding fresh spores in position for next spring. Other root fragments contain both hyphae and vesicles, which can bide the time until plant growth resumes. External hyphae deteriorate once separated from the host plant, whereas hyphae located within roots can remain viable for as much as six months as active propagules in whole form. These mycorrhizal hyphae utilize the starches and sugars of senescent and dead root tissue to overwinter.

Vesicles store lipids, making all such fungal storage units yet another source of carbon energy for eventual regrowth of hyphae.

Pull up garden plants by the roots in the fall, however, and all this changes. There are published reports of root fragments and dried mycelia losing viability after only a few weeks of routine storage. Fungi accordingly rely on spores for the long haul.

Arbuscular mycorrhizal fungi don't produce fruiting bodies like mushrooms, puffballs, and truffles. That more familiar route (to the human eye) belongs to the ectomycorrhizal fungi. Belowground, AM fungi asexually produce spores by various means. Tiny sporocarps, which contain relatively few spores, mimic the cluster theme of a puffball, only on a much reduced scale. Species that create vesicles within plant roots can pack spores away within these cellular compartments. Spores produced at the ends of specialized hyphal strands in the external mycelium seemingly float throughout the root zone. Those spores tucked into intercellular spaces of the root have an "intraradical nature," which lends itself to mass production.

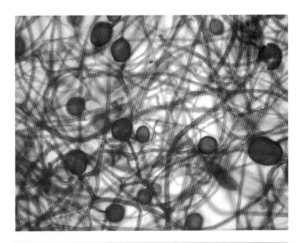

Spores of arbuscular mycorrhizal fungi seemingly float throughout the extraradical mycelium.
Courtesy of Angela Hodge, University of York.

Underground origins do not facilitate widespread spore dispersal. The large size of arbuscular-type spores and thus their weight (compared with spores of mushrooms and puffballs), coupled with characteristically smooth coatings, makes routine windborne dispersal unlikely. Don't rule out a dust twister, of course, but don't wish for one either. Lack of protective pigments renders all AM spores unprotected against the damaging rays of the sun. Finally, these fungal spores are sensitive to desiccation if exposed to the elements over extended periods. Not surprisingly, it's left to small burrowing animals to carry AM spores the furthest afield, by passage through the gut and attached to fur. Spores consumed by earthworms are not actually digested but end up being transported great distances (from the perspective of a microbe) as well.

Spores are packed with carbon nutrition in the form of fatty lipids and carbohydrates to provide for hyphal initiation and growth when the time comes. Germination can get under way in moist environs without plant roots, but really kicks into gear when stimulated by root exudates. The major signaling molecule put forth by roots appears to be a sesquiterpenoid compound, drawing the fledgling hypha in the direction of the root. Limited mycelium branching occurs while the fungus zeroes in on its target.

Nature hedges her bets, nevertheless. This protoplasm investment can be retracted if root presence proves too far away (more distant than a few millimeters), thereby allowing some spores to retain viability for another year. The metabolic rate of spore-initiated hyphae slows down when no actual roots are nearby. This adds another component to the long-term survival strategy of fungi tucked within thick, resistant walls of chitin. Heavy metal concentrations and salinity are additional factors interfering with spore germination.

Spore assemblage of twenty or more different species can be quite common in any one place. Not surprisingly, researchers distinguish these fungi by means of the spores. The International Culture Collection of Arbuscular Mycorrhizal Fungi (INVAM) maintains cultures of recorded species for study and distribution. The section on fungi found on the INVAM website (see resources) informs how these organisms are named and classified, with individual pages documenting properties of all species in the collection.[14]

The importance of ectomycorrhizal spore banks comes fully to light after a forest fire. The living mycelia in such affected ground are seriously set back by high soil temperatures. Tree seedlings return before the fungi necessarily do. Spores from afar must save the day.

This correlates well with the ways of symbiotic basidiomycetes. These fungi sprout fruiting bodies aboveground to disseminate spores. Species in this category produce mushrooms and puffballs, with truffles tucked ever so near to the soil surface. Released spores can handle being exposed to wind, weather, and sunlight. Accordingly, ectomycorrhizal spores are made to "ride the wind" and thereby bring connection back to trees where fungal benefits have been waylaid. EM spores are much smaller than AM spores and often have spines or other surface features to give more lift in the wind. These ecto-type spores are typically pigmented so as to withstand UV radiation as well. Finally, most of these propagules are very dry so as to resist desiccation until conditions are right for germination. Rain carries these tiny spores into the soil to seek a rooted home.

All spores require a period of time to colonize roots, on the order of three to six weeks. Germination, responding to root signals, and then locating the source of that allure takes what time it takes. Keep in mind that root fragment inoculum can do this within

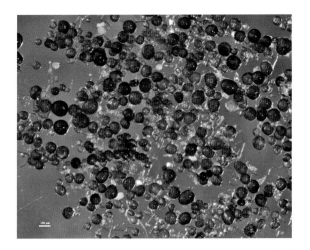
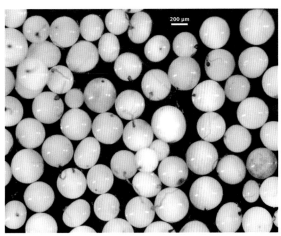

Left, spores of *Glomus deserticola* and, *right*, spores of *Gigaspora margarita*. Courtesy of International Culture Collection of Arbuscular Mycorrhizal Fungi, West Virginia University.

The spores of *Scleroderma citrinum*, more commonly known as the poisonous pigskin puffball, will mature by the time the warted peridium cracks open on its own accord. Photo by David Spahr.

as little as a week due to the presence of waiting hyphae. This distinction can have relevance in the greenhouse, where chopped roots can outpace spore inoculum in getting seedlings off to a head start.

Another practical application of the spore time line worth noting concerns how long it takes for these assorted fungi to form spores. No individual fungus lives more than several months to a year, and this is little more than a guess on the part of researchers, given the complexity of mycorrhizal networking. (What could be called the *fungal being* is another matter entirely.) We'll frame "fungal maturity" as the point where the reproductive phase begins, and do so in the context of a temperate growing season, as that's where this bit is heading. Asexual spores are produced by budding in the case of arbuscular mycorrhizal fungi. Formation of spores takes place approximately three to four months following root colonization by most species. The relevance of this span of time will come up when we consider carrying fungal potential forward by means of cover cropping.

Nutrient Dynamics

Mycorrhizal fungi acquire nutrients from decomposing plant litter and soil pools. These can be anything from exchangeable cations and anions found on soil particles to veins of fractured rock. The importance of each of these pools varies greatly by element. Carbon and nutrients can also be exchanged from one plant to another via mycorrhizal networks, which we'll explore further on.

Free trade between fungi and plants starts with straightforward economic principles. Plants have photosynthate sugars to offer to mycorrhizal fungi, which otherwise can't access carbon. Fungi in turn assist the plant by facilitating the uptake of mineral nutrients and water. The resulting exchange of nutrients provides opportunities for each symbiont to decide where the better deal of the moment can be found. One impressive ramification of this underground economy from the healthy plant perspective is that plants receive a balanced array of nutrients (see "Balanced Nutrition" on page 59) by means of biological barter. Recognizing which nutrients mycorrhizae can deliver and how that's accomplished establishes the ground rules.

Mycorrhizal fungi secure nutrients by secreting digestive enzymes wherever hyphae lead. We'll learn shortly that bacteria often have a role in dissolving the targeted substrate as well. Substances are broken down into elements. Simple nutrients are then absorbed through pores in hyphal cell membranes and carried along to roots. Functional exchange with the plant host takes place across the intraradical interface, be it intercellular spaces breached by ectomycorrhizae from without or the arbuscular structure of endomycorrhizae within the root cell itself.[15] Plants put forward organic carbon at these shared boundaries . . . which gets absorbed by the fungus . . . and then is immediately transported in the form of converted sugars and sugar alcohols to the rest of the fungal organism.[16] The root in turn absorbs its end of the deal. Decisions are made by means of diffusion gradients (the movement of elements from a high concentration to a low concentration), but whether the fungus or the plant is "in charge" is purely speculative.

Researchers list phosphorus and nitrogen as the principal nutrients delivered by mycorrhizae to plants, followed by calcium in depleted soils. Hyphae pass along sulfate released from organic matter. Important side dishes include iron, copper, and zinc. Certain mycorrhizae are noted for increasing levels of potassium, magnesium, or manganese. Every pairing of fungus and plant brings its own adaptation to the nutrient scene — yet relatively few absolute affiliations have been corroborated by research to date. And that's okay, when untold diversity assures that many different players will always be in the game.

Phosphorus concentration has been shown to increase up to four times in mycorrhizal plants. The increased outreach of extraradical mycelia beyond the spread of roots and fine root hairs explains this in part.[17] Soluble phosphorus travels to the roots via diffusion, yet quickly runs out in terms of direct contact. Roots are unable to access more phosphorus without further extension beyond the developing depletion zone. That requires additional energy, whereas mycorrhizal hyphae access considerably more soil volume as far as the fetching of available phosphorus goes. Still, this macronutrient is rather immobile (see "Phosphorus Addendum," page 109), and thus the supply on hand constantly needs a reboot. Enter the phosphate-solubilizing bacteria, which go to work on apatite (calcium phosphate) and other phosphate-rich rock.

Fungi pay these bacteria with hard-won carbon to further the cause. Organic acids produced by phosphate-solubilizing bacteria chelate the cations that bind negatively charged ions of phosphate, thereby releasing more phosphorus in a soluble form. Humic substances are involved as well when organic matter is plentiful, coating aluminum and iron oxides, which reduces the amount of phosphorus seeking to bind anew with positively charged soil particles. This ability of fungal hyphae to interact with other microorganisms opens up truly hardcore avenues for nutrient exploration.

Nothing typifies this more than the "bacterial bore" found on hyphal tips bound for glory in mineralized rock. Ectomycorrhizal hyphae range in diameter from 2 to 20 micrometers, right on the edge of resolution for human eyesight. This fineness allows penetration into untapped nutrient zones where roots and even root hairs prove too coarse. Amazingly, hyphae will even commence to drill directly into solid rock to open up biologically induced fissures that over time will lead to further fracturing of the bedrock.

This microbial weathering actually involves two enzyme systems at work. Release of fungal exudates is concentrated at the tips of active hyphae. Much of this will be consumed by bacterial communities working slightly back from the point of hyphal impact. High carbon levels in turn stimulate the production of organic acids by the bacteria . . . which hew apart the rock substrate grain by grain . . . freeing potassium, calcium, and magnesium ions from the mineral. Kind of like a rotary hammer drill being powered by a microbial fuel cell. Fungal tunnels widen and deepen as untapped mineral wealth becomes available directly to plant roots.[18] Research has revealed that at least half of the ectomycorrhizal fungal taxa found in the bony soil of northern Sweden were exclusively associated with mineral soil horizons.[19]

Mycorrhizal hyphae of all persuasions are able to absorb inorganic nitrogen in its ammonium form. Nitrate ions are typically taken up just as readily, except by certain ectomycorrhizal species in boreal forests. And things get all the more interesting when we consider organic nitrogen prospects.

Amino acids in soils exist in amounts that could make a considerable contribution to mycorrhizal uptake of nitrogen. Plants that obtain nutrition in ever more complex forms beyond the ion gain a metabolic advantage (see "Biological Reserves," page 37), which makes this avenue of inquiry all the more pertinent. It's well known that ectomycorrhizal fungi utilize amino acids and peptides

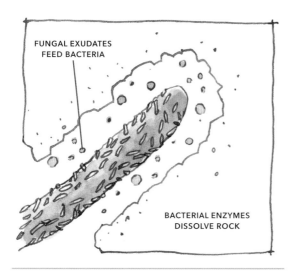

FUNGAL EXUDATES
FEED BACTERIA

BACTERIAL ENZYMES
DISSOLVE ROCK

Bacteria work in concert with fungal hyphae to bore into solid rock to release minerals by means of enzymes and organic acids.

(partial proteins) in forest ecosystems. Ericoid mycorrhizae actually have few alternatives other than organic nitrogen in the acidic habitat typical of these associations.[20] Pollen, nematodes, springtails, and saprotrophic mycelium are often "it" for biologically relevant substrates. Arbuscular mycorrhizae have a strong tendency to correlate with nodulating bacteria in passing along a nitrogen fix through the mycelial network. While the capabilities of mycorrhizal fungi to extract carbon from plant litter are a big fat zero, this does not rule out nitrogen release and absorption directly from decomposing organic matter. Arbuscular-type fungi typically deliver arginine (one of the standard amino acids) to the intraradical mycelium. This gets broken down to provide ammonium to

The ability of organisms to obtain nutrients from rock can readily be seen wherever lichens take hold. This form of mutualism between two species consists of fungal filaments surrounding cells of green algae or blue-green cyanobacteria.

the plant, while the other molecular bits apparently remain in fungal tissue.

Overall, the nitrogen contribution from mycorrhizal fungi can be held just as often in reserve. Bacteria break down amino acids in the rhizosphere (root zone), which provides for direct root absorption. Fungal hyphae reaching further afield will uptake nitrogen to deliver to the plant only when nearby supplies are limited, organic or otherwise.[21] Mycorrhizal involvement keys to whether a soil is nitrogen-impoverished or laced with the stuff. That's determined by a robust cycle of assimilation and mineralization by all sorts of organisms working in concert.

Hyphal Lysis

Molecular insights into more complex variations on nutrient flow are being revealed at a fast pace. Yet there's a flip side to mycorrhizal nutrient delivery that's too often overlooked. Like the swallow song of summer, all good times eventually come to an end. Each round of hyphal investment has a fixed life span of varying degrees. How this will tie into healthy plant metabolism in the next chapter should absolutely flip your gourd. Let's start by establishing the concept of *hyphal lysis*.

The disintegration of a cell and the subsequent release of protoplasm describe the end of time for a fungal hypha. This is lysis, involuntary cell death, resulting from the rupture of the cell wall after being cut off from adjacent hyphae. Consider the abandoned arbuscule, having served as a productive intraradical interface for the allotted seven days (give or take), closing down in what's in truth a stressful environment.[22] It is inside the root cell, awash in extracellular fluids, stretching the cell

membrane almost unbearably. Something has got to give. Each degenerating arbuscule gets shut off by a cross wall, thus preventing backflow of nutrients. *Stay focused now.* Complex fungal nutrients are being released directly into plant fluids. *Meaning what exactly?* Phantasmagoria!

Lysis results in the release of structural cell wall proteins, acidic polysaccharides, lipids, and polyphosphates. High concentrations of calcium are made available. The fungus literally gives its all to the plant. Nutrient flow via extracellular fluids will go in numerous directions — new root ventures, rich exudates to draw more microbes, upward into leaves and shoots, the next fungal bay over, perhaps even into that nutrient-dense apple. The resulting energy gained by the plant in turn makes possible greater investment in immune function by the plant.

Lipids formed through the conversion of glucose to triglycerides represent stored energy for any lifeform. Much ado should be made about fats. Japanese researchers did precisely that in 2014 by revealing the first direct link between arbuscule collapse and the streaming of lipid droplets into the root cortex.[23] The emerging lipids in all likelihood were preformed and held within the arbuscular trunk rather than being integrated on the spot. Lipid synthesis in mycorrhizal fungi involves interaction with the fatty nature of the root cell membrane along with "bidirectional streaming" of protoplasm from arbuscule to intraradical hyphae to extraradical mycelium and back. That some of the good stuff gets shared with plant hosts when the arbuscule gate shuts at the time of lysis makes complete sense.

And yet this runs counter to what scientists thought even a decade ago.[24] The pace of knowing on our part does not alter how creation works. New research like this simply gives us another piece of the puzzle to contemplate in the fascinating montage that makes life possible.

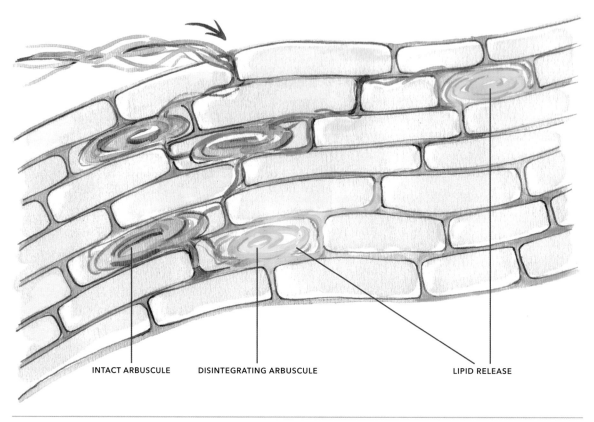

INTACT ARBUSCULE   DISINTEGRATING ARBUSCULE   LIPID RELEASE

The collapse of arbuscules within root cells is one way by which mycorrhizal fungi contribute "partially built nutrition" to robust plants. Microscopic photography (with selective staining of fungal structures and live imaging) has revealed that droplets of lipids are released into the root cortex following lysis. Adapted from Y. Kobae et al., "Lipid Droplets of Arbuscular Mycorrhizal Fungi Emerge in Concert with Arbuscule Collapse," *Plant Cell Physiology* 55 (2014): 1945–1953.

Nor does the extraradical mycelium last forever. The law of ebb and flow mandates that all parts of a fungal organism come available in due time. The ongoing spectacle of microbe consuming microbe offers no assurance to mycorrhizal fungi of a species pass. Actinomycetes in particular are quite capable of decomposing chitin, the structural component of fungal cell walls. Perennial plants going into dormancy at year's end experience an autopilot version of nutrient release. Root systems enter a sleepier phase. Fungi go into spore mode and hide away as hyphal fragments in roots and crevice places. Mycelial structure dissolves; a portion gets taken in by roots in more complex forms; the bulk assimilates into the jaws of the soil food web. Fungal death again contributes carbon and elemental nutrients in high concentration.

CHAPTER 2

Healthy Plant Metabolism

The existential connection between mycorrhizal fungi and healthy plant partners is indisputable. What we need to understand next is what it means to truly be a healthy plant:

To have leaves out in the bright sunshine
To sway with the breeze yet stay firmly rooted
To dance with microbes
To flower, to fruit, to seed
To write sonnets in green

Healthy plant metabolism begins with a molecule of water, a breath of carbon, and light energy from our nearest star. The tangible science behind all this unlocks the righteous way to farm and garden, give honor to trees, and plain do right by this earth. Nothing has ever excited me more.

A few core concepts set the stage for this exploration into plant health.

The dictionary tells us that *health* reflects the functional efficiency of a living organism. Primary metabolic processes for plants involve photosynthesis, respiration, and the synthesis of organic compounds needed to sustain life. Being sustained sounds kind of good, like bearing up under stress. Wherein enters the term *plant secondary metabolism*. These phytochemical pathways aid in the growth and development of plants, but are "not required for the plant to survive." Hmmm. That parting shot is taught the world over in biology classes (and thus the quotes), yet what a dumb thing to say to a grower or an herbalist.

Reductionist science has gotten us into trouble before. The ensuing biochemistries of vascular metabolism play a pinnacle role in keeping plants healthy. A wide range of constituents are used as defensive mechanisms to ward off herbivores, pests, and pathogens. Some metabolites make leaves taste bitter, and can even prove toxic. Most notably, quite a number of these compounds not only trigger a resistance response, but become the means by which plants thwart fungal and bacterial disease. Plant hormones regulate metabolic activity within cells. Allelopathic interactions inhibit the growth of competitors. Pigments such as carotenes and flavonoids color flowers and, together with phenolic odors, attract pollinators. Another phenolic compound known as *lignin* adds stiffness and strength to cell walls. This complex polymer provides structural support to trees reaching upward to the sky as well as amber waves of grain rippling in the breeze.

And yet, all such secondary metabolites are viewed as nonessential to the functioning of plants?[1] Sounds like a plot to promote human intervention to me. Serious kidding aside, let's ponder for a minute what it means to be an unhealthy plant.

A number of things may go awry. Aphids crowd in to suck plant juices. Moth larvae nip and tuck at new growth. Scab, rust, and blight spread mercilessly. Beetles skeletonize leaves till only veins remain. Mineral deficiencies abound. Bacterial canker and eye of newt. Rotting root and cauldron bubble. *Something wicked this way comes!*

Modern farming systems are regrettably unhealthy by choice.[2] Fungicides, insecticides, and herbicides are no more than medications applied to compensate for practices that harm soil life. The resulting harvest may be "productive," but the food itself lacks meaningful nourishment by half, and offers even less in the way of antioxidants needed by our bodies to ward off degenerative disease. On the other hand, dealing with countless symptoms spurred on by empty foods has become quite the boon for modern medicine. All comes round when we realize that parallels exist between human health and plant health.

Pests and pathogens encountered by plants should be considered "symptoms" resulting from a breakdown of natural defenses. Just as we humans have an immune system to ward off everyday encounters with germ organisms, plants have phytochemical abilities with which to face similar encounters in the green world. Plant resistance to insect feeding depends upon metabolic processes going the distance . . . along with an assist from beneficial insects. Plant resistance to pathogenic fungi and bacteria depends upon metabolic processes going the distance . . . along with competitive colonization on the part of friendly microorganisms from root tip to shoot tip. The prospects for such complete metabolism, of course, pivot on mineral availability and mycorrhizal collaboration.

Plants with full access to balanced nutrition provide a very different bill of fare for pests and pathogens seeking a niche to exploit. Our foray into a health perspective will assume a wide range of minerals are on hand in reasonable proportion. Later on, we will put on our thinking caps to explore just what proper mineral investment might look like. Mycorrhizal fungi perform an important role in balanced nutrition as well. Select nutrients go to the host plant offering the better carbon trade. Sometimes the deal involves partially built proteins; other times it's trace mineral complexes from newly mined rock. The abundance provided every time one organism gives way to another frees up the profoundest offerings of all. This underground economy invokes ecosystem intelligence and an interdependence of species that we'll be marveling at shortly. Plants in turn take these elements and provide oxygen and sustenance for all who walk in beauty on this good earth.

We'll proceed under the unified banner of healthy plant metabolism. From here on in let's think of plant health in the affirmative. *Vibrancy. Vitality. Joie de vivre.* These are the words to seize the day and set our course aright.

Phytochemical Progression

Sunlight captured by a leaf cell miraculously transmutes carbon dioxide and water into a simple sugar. The resulting carbohydrates are used by the plant in turn to trade for minerals down in the root zone. Other sugar molecules are joined with nitrogen to form proteins. Lipids (fats) are created to store accumulating energy. This cascade of synthesis ultimately leads to the production of resistance metabolites, which propel immune function in a healthy plant.

What sounds relatively simple depends on a number of factors that aren't necessarily in place in all soils. Complete carbohydrates, proteins, and lipids are only formed when the enzyme systems of plants run smoothly. That in turn depends on a full complement of trace minerals being available. These processes require a certain level of efficiency for plants to have "energy enough" to fuel natural plant defenses. That oomph results from teamwork within the soil food web, meaning mycorrhizal fungi and all the other players involved with nutrient transfer.

A French agronomist named Francis Chaboussou asked a fundamental question some fifty years ago: What is going on, chemically speaking, for a healthy plant versus an unhealthy plant? He documented a vast amount of research over decades on protein formation in plants and how this correlated to a plant's chances of either attracting or resisting insects and disease. Chaboussou went on to develop the theory of trophobiosis, whereby he explained what happens when plant metabolism goes awry. In a nutshell, a deficiency of minerals along with biological imbalance leads to an inhibition of protein synthesis in plants. He indicted agrochemicals for causing such disparities.[3] The resulting accumulation of soluble substances in plant sap provides improved nutrition for pests and rapid multiplication and virulence of bacterial and fungal pathogens. The clarity of this proposition was dismissed by his peers precisely because it challenged too many preconceived notions.

We know a thing or two about human progression as well when it comes to understanding natural systems. A seemingly objective assumption establishes how things have always been done. Case in point, synthetic fertilizers promoted as a surer means of crop production than a diverse soil biota. Were we still able to ask Sir Albert Howard about this (whose writings in the previous century urged on the organic movement), he would tell us to forgo that NPK and apply more compost.[4] Our grasp of what-seems-to-be-definitive science continually shifts as we advance in fits and starts. Someone new arrives, ponders the moment, and then applies unbounded curiosity and intuition in asking even better questions. Why obsess on disease when we might work directly to facilitate plant health?

Enter John Kempf, a young Amish farmer from Ohio, who sought answers to move beyond a series of crop failures on his family's farm. His "aha moment" came when learning about plant immune systems through self-directed study. (A credible education, indeed.) The solution lay with balanced nutrition to the nth degree. Plant sap analysis helped John discover specific correlations between trace minerals and the different phases of phytochemical progression.[5] Add a touch of remedial biology and growers have exacting tools to further the work begun by Chaboussou.

Only our efforts will take place on hallowed ground from here on in. No agrochemicals. No synthetic fertilizers. No messing with healthy plant metabolism. Those free amino acids and reducing sugars in a plant's cell sap that feed pests and disease can be brought into the fold of effective protein synthesis from the get-go. Further health ramifications lie beyond, and it's the biology that holds that trump card.

MYCORRHIZAL PLANET

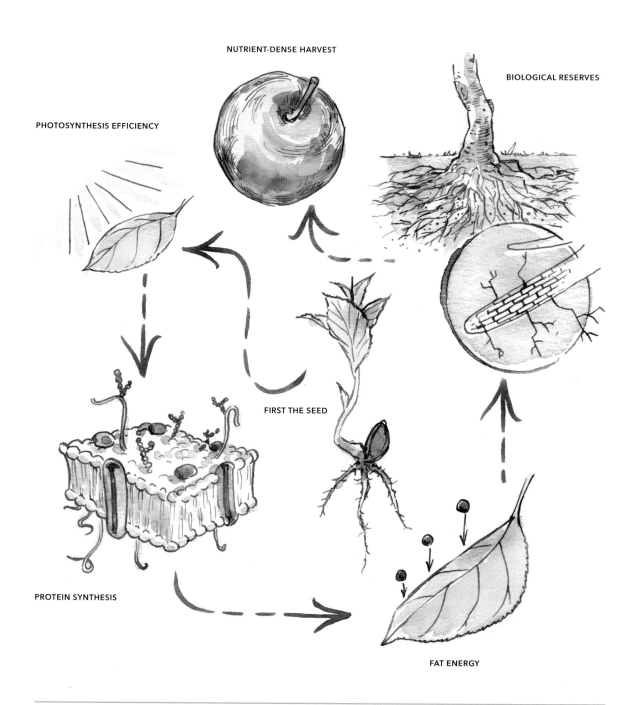

NUTRIENT-DENSE HARVEST

BIOLOGICAL RESERVES

PHOTOSYNTHESIS EFFICIENCY

FIRST THE SEED

PROTEIN SYNTHESIS

FAT ENERGY

Phytochemical progression kicks into gear with efficient photosynthesis brought about by full availability of trace minerals. Complete protein and lipid synthesis follows from there. Resistance metabolites, kindled in part by mycorrhizal fungi, round out natural plant defenses.

PHOTOSYNTHESIS EFFICIENCY

Plant metabolic processes commence with the production of glucose through photosynthesis. Leaf cells contain *chloroplasts* that act as photoreceptors to trap sunlight. These organelles may very well be photosynthetic bacteria that adapted to life within the earliest plants.[6] The green pigments found in chloroplasts absorb the sun's radiation, thereby discharging high-energy particles. Electrons thus freed from the chlorophyll fuel the transformation of carbon dioxide and water into a simple sugar and oxygen.

Photosynthesis in green leaves the world over is the everyday miracle that makes our lives possible. The air that we breathe. The food that propels our every step. All because fungi teamed up with plants and a few cyanobacteria in the predawn hours of now.

Glucose, as one of the simplest of sugars, cannot be hydrolyzed to smaller carbohydrates. All carbohydrates build on from glucose with just three elements involved: carbon, hydrogen, and oxygen. Other monosaccharides include fructose and xylose. These simple sugars combine to form disaccharides such as sucrose and maltose. Yet more complex sugars called polysaccharides develop as additional carbon atoms bond on for the ride. These are giant molecules that, importantly, are too big to escape from the plant cell. Starches, glycogen, and cellulose can be broken down by hydrolysis into monosaccharides when energy is needed by the cell. Welcome to the first phase of plant health.

Dynamic plants store an increasing portion of carbohydrate reserves in more complete forms. Internal resistance to soilborne pathogens follows from there. Diseases caused by take-all fungus in wheat and root-knot nematodes result in part from an excess of simple carbohydrate sugars. The significance of "what's for dinner" will prove strategic aboveground as well. Consider a human analogy. Sweets that prove addictive to our less-than-prudent selves engender the right substrate for germs in our bodies. This makes sense likewise for plants.

An efficient photosynthesis process nudges on not only carbohydrate complexity but subsequent phytochemical progression as well. The four principal groupings of plant compounds are carbohydrates, proteins, lipids, and the resistance metabolites (including essential oils). These organic macromolecules are created through *biosynthesis* involving numerous steps. Enzymes spur on the action as plant constituents are modified, converted into other compounds, then joined together to form long chain polymers. Carbohydrate synthesis begins with glucose. Protein formation keys to amino acids brought about by the incorporation of nitrogen. The underpinnings of lipids are fatty acids. All come versed in carbon. All require an ionic assist.

Enzymes increase the rate of reactions by lowering the energy threshold for activation to occur. These catalyst proteins attach to one compound and then, zip . . . the subsequent compound exists. This is what makes photosynthesis "efficient" with respect to plant defenses. We're going to skip a lot of big words now (like aspartate aminotransferase) and go straight to the takeaway. Metal ions are the common *cofactors* required by enzymes to keep things humming.[7] The principal minerals involved are iron, manganese, cobalt, copper, zinc, and molybdenum.

Making sure the full range of trace minerals is available at critical points in crop growth cycles promotes green health. Biologically robust soils go a long way toward providing this. Yet balanced

Cofactors of Plant Enzymes

Iron (Fe) is required for chlorophyll synthesis as well as in many oxidation/reduction reactions. Medium pH is critical for iron to work effectively as a cofactor.

Manganese (Mn) serves as a cofactor in twenty-four enzyme reactions. The assimilation of carbon dioxide in plant respiration depends on it. Manganese takes on important roles in the metabolism of nitrogen and fatty acids.

Cobalt (Co) gets an "insubstantial rap" by plant physiologists in terms of being essential. Yet eight enzyme systems rely on cobalt as a cofactor. Methionine synthase in particular is involved in complete protein production.

Copper (Cu) functions primarily as a cofactor for a variety of oxidative enzymes. Copper is necessary for proper photosynthesis and subsequent production of carbohydrates.

Zinc (Zn) is absorbed as a divalent cation by plants. Twenty-five enzyme systems rely on zinc as a cofactor. Zinc partners with copper to produce a potent antioxidant enzyme that protects plant cell walls.

Molybdenum (Mo) is a cofactor with two enzymes involved in transforming nitrate to ammonium. This places the "tongue-twister mineral" front and center in building amino acids through nitrogen metabolism.

Nor is this solely about the micronutrients. **Calcium (Ca)** functions as a cofactor with many enzymes and is particularly important in cell wall synthesis. **Magnesium (Mg)** is critical for enzyme functioning and an integral component of the chlorophyll molecule. **Sulfur (S)** is important in both protein structure and fatty acid metabolism. Additionally, sulfur works with iron to facilitate electron transfer in photosynthesis.

nutrition for crop plants often calls for a bit more daring. Foliar applications of trace minerals to promote complete biosynthesis are gladly made when you keep the alternative in view.

PROTEIN SYNTHESIS

The second phase of plant health involves amino acids brought together to form proteins. Many insects lack the necessary enzymes to digest complete proteins. Similarly, those fungal pathogens that seek sustenance directly from the plant cell do so by absorbing soluble amino acids. Those two factors combined are more than enough to shift crop expectations in the right direction, so let's follow this trail through to the end.

A protein is composed of chains of amino acids linked by peptide bonds. Every synthesis reaction between a pair of amino acids results in the release of a molecule of water. There are hundreds of amino acids found in plants, of which twenty are recognized as the "standard amino acids" for building complete proteins. An inclusive protein has a stable conformation, whereas a less complete protein (known as a *peptide*) generally refers to a short amino acid oligomer often lacking a stable three-dimensional structure.[8]

Protein formation incorporates another essential element into the life equation. Soluble sugars, when partnered with nitrogen in its ammonium form, are the base materials used to form amino acids. Obtaining nitrogen is a multistep process for plants. Nitrogen gas in the air needs to be fixed by certain bacteria in the soil. Other organisms release "inorganic nitrogen" when consuming one another.[9] Plants in turn can absorb nitrate or ammonium from the soil via their root hairs. If nitrate is absorbed, it first needs to be reduced to ammonium ions for incorporation into amino acids, nucleic acids, and chlorophyll.

The pace at which amino acids become complete proteins ties to the availability of trace minerals. We're back on familiar ground in recognizing the role of enzymes in facilitating metabolic activation. Meanwhile, with photosynthesis ramped up, plants transfer greater quantities of sugars to the root system and the surrounding microbial community. This stimulates the release of minerals (along with nitrogen) as microbe upon microbe responds to the increased carbohydrate energy. Mycorrhizal fungi take note as well, delivering trace minerals from the soil matrix in plant-available form in exchange for that carbon sugar. Plants then utilize these metal ions as cofactors to stimulate the enzyme scene so needed to form complete carbohydrates and especially proteins. Talk about compounding interest on investment!

Effective protein synthesis results when we respect natural systems. We do that by minimizing harm to soil life and supplementing mineral nutrition. Easy enough. Growers will recognize this second phase of plant health as the point where the real fun begins.

Pest encounters of the protein kind work like this. The lives of insects verge on frenzy. Large amounts of protein are needed to fulfill a fast-paced molting cycle, from eggs to larvae to pupae to adults. Yet many species of insects lack the digestive enzymes needed to break down complete proteins. Let's ponder the implications of this.

Plants that are constantly forming more inclusive proteins are not attractive to insects with a simple digestive system. Conversely, insects will source incomplete proteins from plants having elevated levels of soluble amino acids in the plant sap. And just who might these problematic insects be? Sucking insects, from mites and aphids to white flies and leafhoppers, most certainly. Leaf rollers in orchards, green cabbage worms and corn borers, even those nasty tomato hornworms find little joy in a high-protein diet. Larvae of alfalfa weevil and currant sawfly simply say no thanks to healthy offerings.

So shall we apply minerals or pesticides? Seems like a no-brainer to me.

Certain pathogenic fungi establish an ongoing feeding relationship with the living cells of a host plant rather than killing those cells outright. Such *biotrophic* organisms cause mildew and rust diseases. Other leaf spot diseases deploy the same strategy in the early stages.[10] These assorted ruffians invade host cells to produce nutrient-absorbing structures, which in turn support substantial infection lesions within the subcuticular layers of the leaf. This type of fungal parasitism results in serious economic losses by reducing the competitive abilities of crop plants.

Growers can unmask biotrophic-type infections by fully understanding nutrient sink dynamics. Disease fungi require both carbon and nitrogen to establish and maintain the infection site. Soluble amino acids in plant sap offer both elements. Let's zoom in to "see" what happens next. A pathogen spore must find its way to these food resources to get the ball rolling. Penetration hyphae either wriggle through stomata (respiration openings) on the surface of the leaf or punch directly through the waxy cuticle and the plant cell wall. The hyphal peg (tip) must withstand plant defenses while at the same time stimulating the release of plant nutrients for its own use.[11] Time for the *haustoria maneuver*.[12] A nutrient-absorbing haustorium forms on the end of each hyphal peg once inside the cell wall. These branching structures effectively increase the surface area in contact with the semipermeable cell membrane.[13] Soluble amino acids and simple sugars are awash in the extracellular fluid between the cell wall and the cell membrane. Pathogen protoplasm oozes out through minuscule pores in haustoria walls and intersperses with the cytoplasm of the host cells. Deals are made.[14]

Or maybe not. The supply of amino acids ebbs and flows as enzyme action upgrades peptides into more complex proteins that readily embed in the

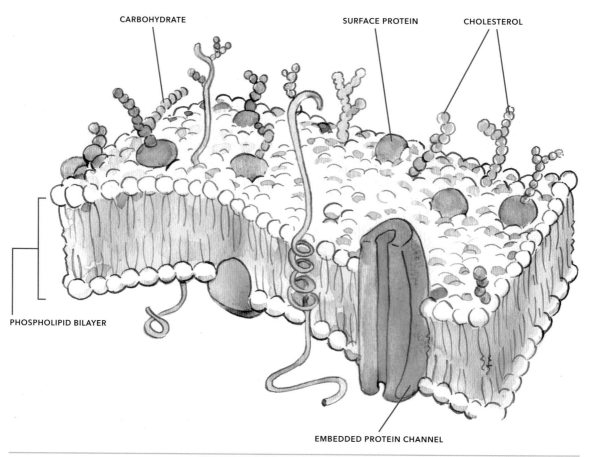

CARBOHYDRATE   SURFACE PROTEIN   CHOLESTEROL

PHOSPHOLIPID BILAYER

EMBEDDED PROTEIN CHANNEL

The plant cell membrane consists of a phospholipid bilayer with embedded protein channels and surface proteins.

cell membrane. Once again it's time to revel in trace mineral cofactors.

One stellar example of all this is the link between cobalt and apple scab disease. Pathogenic fungi have preferred foods, and in the case of *Venturia inaequalis*, several of the standard amino acids found in leaf cells more than qualify.[15] Cobalt is a cofactor in incorporating amino acids into a complete protein. Including this trace mineral in orchard foliar applications in spring when scab is on the prowl takes more amino acids out of the food stream for this disease organism.

Immune function applied; no damage done.

Fat Energy

All pathogenic fungi struggle to gain infection footholds on plants brimming with fatty acid compounds. The third phase of plant health kicks in as photosynthesis efficiency and effective protein synthesis enable plants to store more energy in the form of lipids in both vegetative and reproductive tissue.

Plant fats take many forms. Cutins and waxes mediate between the air and the plant's epidermal cells. The principal function of the extracellular cuticle layer is to prevent evaporation of water from leaves and shoots. These protective lipid

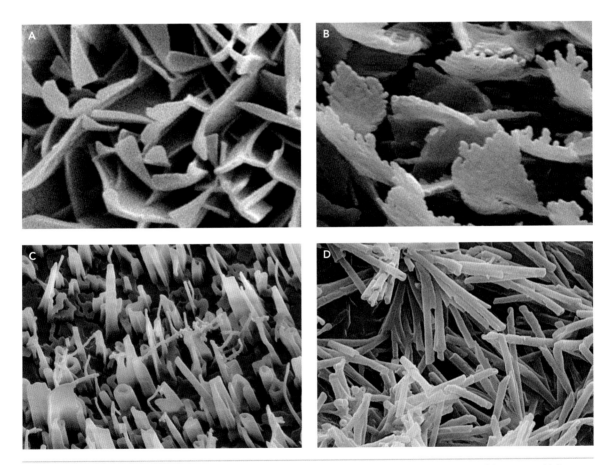

Epicuticular waxes on the outer cuticle surface on (*a*) myrtle spurge, (*b*) yucca, (*c*) wild cabbage, and (*d*) eucalyptus. Photos by Hans Ensikat, Nees Institute for Biodiversity of Plants, University Bonn.

compounds are generated from the polysaccharide matrix (cellulose, hemicellulose, and pectin) composing the primary cell wall. Phospholipids form the cell membrane that lies just behind the cell wall. This lipid bilayer allows for the passage of fatty acids but blocks the movement of water-soluble molecules.[16] Storage lipids accumulate as triglycerides in seed oils and to a slight degree within leaves.

Plant metabolism commits to baseline levels of lipids for these functions even in the worst of times. Yet lipid investment will be raised as much as three to four times when phytochemical progression moves along at a healthy clip. The cell membrane becomes stronger, the cuticle ever more resilient.

The *cuticle defense* to airborne fungal pathogens consists of several legs. The natural advantages result from an engaged soil biology and mineral investment. Boosting lipid levels yet again for crop plants can be done with "fatty acid sprays" at critical junctures in the growing season. (See "Fatty Acids," page 74, to get started, with targeted advice to follow in chapter 6.) Fermented plant extracts augment this scene in specific ways. Arboreal microbes make things highly competitive . . . and

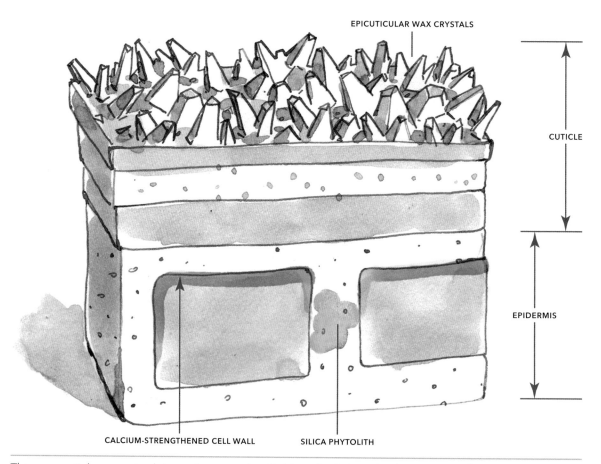

EPICUTICULAR WAX CRYSTALS

CUTICLE

EPIDERMIS

CALCIUM-STRENGTHENED CELL WALL

SILICA PHYTOLITH

The waxy cuticle serves to deter pathogens when lipid synthesis is optimal. Boosting cell wall strength with calcium and blocking intracellular spaces with silica only makes the "cuticle defense" all the more formidable.

these allies use applied fats to maintain that vigil. Lipids sustain good health across the board.

Let's zone in further. Fungal spores are dispersed by wind and rain. This airborne access for disease organisms depends upon the plant surface staying moist. Hyphae sent forth from the germinating spore rely on enzymes to penetrate the waxy cuticle and cell wall.[17] Hindering this process in any way means the spore may not have enough time to accomplish its mission. Friendly organisms on the surface of the plant secrete antimicrobial substances to stop the invader. Resistance metabolites of the plant's own devising put on the brakes. Translocation of calcium strengthens the cell wall within. Silica adds a "slippery slope" factor to cuticle dynamics.[18] A more acidic moisture puddle on the leaf surface slows down the pace of enzymatic chemistry.[19] But mostly . . . it's the waxy, impassable, ornery ol' cuticle itself that gets the job done.

Certain fungi invade plant tissues aggressively, killing host cells immediately to obtain nutrients. Such *necrotrophic* organisms cause rots and molds on fruits and vegetables. These diseases often enter the crop through cracks in a weak cuticle or sting wounds from insects. Necrotrophs do not produce specialized penetration structures, but rather rely on secreting toxins to degrade the cell wall beyond the cuticle. Once inside, rots readily declare *game, set, match*.

Only, this will not happen if the cuticle is strong and insects aren't drawn to feed on incomplete proteins in the first place. Phytochemical progression is a complete package.

Nor should we overlook that biosynthesis has a reverse gear. Elevated lipid levels serve as provision for rainy days. Stored oils can be converted back to fatty acids and protein catalysts . . . thereby utilizing the energy within to maintain healthy metabolism during extended periods of limited photosynthesis.[20] Fat reserves in plants counter what is known as *proteolysis*, the breakdown of protein metabolism for lack of amino acid building blocks. Stretch your minds around that!

Plants face a sizable cast of airborne ruffians. Despite it all, we still have this marvelous green world to call home. Nature counters the disease realm again and again when healthy plant metabolism is the order of the day. Those plants that achieve higher lipid levels and stronger cell membranes are more resistant to even the most notorious fungal pathogens. Bacterial invaders are another category entirely, but there's one more phase to this zone defense of ours yet to put into play.

BIOLOGICAL RESERVES

Going the distance conveys that plants have "something more" in reserve. The fourth phase of plant health encompasses three large chemical classes — terpenoids, phenolics, and alkaloids — which offer numerous plant defense mechanisms. Antifeedants make leaves less than palatable for herbivores and higher-order insects. Certain hormones help plants cope with a plethora of environmental stresses.[21] Red-hued anthocyanins act as a sunscreen for juvenile leaves not quite ready for prime time. Resistance compounds that function as localized antibiotics and enzyme inhibitors put the ax to persistent pathogens.

This subsequent metabolism comes about wherever plants are fully committed to their microbial friends.

Yet that statement doesn't quite cut it, does it? Of course plants are fully committed! The dualistic nature shared between the very first terrestrial plants and mycorrhizal fungi is as real today as ever, only amplified a gazillion-fold. Human perception of healthy plant metabolism comes full circle only when we acknowledge that what began with symbiosis carries forward with symbiosis. There is no separation between plant and fungus and all the other players in the soil food web. Life's dance with the mycelium encompasses complete carbohydrates, proteins, lipids, and resistance metabolites, because that's how Nature designed things to be. ♪*Sforzando*

Lord knows we lose sight of the most basic connections. The final phase of plant health is brilliantly simple. Its message has been honed especially for us, the growers and landscape stewards and foresters of these times. Don't mess with the fungi, people. Don't think for a minute that soil chemistry somehow trumps nutrient uptake from sister organisms. Do not disturb the impeccable.

Biological alchemy happens because fungi make it so. That reserve energy needed by plants to cross the finish line comes about because mycorrhizal fungi and other soil organisms in the vicinity of feeder roots deliver nutrients in partially built form.

Plants have two basic ways by which to obtain nutrients. Essential mineral elements that are absorbed directly by plant roots exist in the soil solution as ions. Soluble ions move from bulk soil through the rhizosphere before entering roots by means of capillary water flow. Hydroponic growing systems and NPK-driven agriculture emulate this approach to a T. Soluble ions are totally legit nutrients for plants . . . when taken in moderation.

Organism-derived metabolites are the real "health foods" for plants. Choose the analogy that works for you: biodynamic grains, raw milk, organic veggies, grass-fed beef, holistic apples, real olive oil, herbal tonics, or grandma's chicken soup. We can even throw in protein shakes for bodybuilders. The point is, when plants begin absorbing the greater portion of their nutrition as microbial metabolites, the "energy efficiency quotient" of biosynthesis goes through the roof. The energy of the sun that triggers photosynthesis day after day drives metabolic processes. This unleashing energy is finite (in a sense) as glucose converts to sucrose and ultimately to cell compounds infinitum.[22] Plants that conserve energy every step of the way have greater ability to make more immune compounds.

Partially built nutrition in the form of amino acids and chelated mineral complexes delivered by mycorrhizal networks contributes to that "something more" for plants to reach the summit of phytochemical progression. Uptake of nitrogen in its ammonium form inherently saves energy . . . which happens often in a rhizosphere with significant fungal biomass. The resulting "localized acidity" limits the number of nitrifying bacteria that would otherwise convert the ammonium to nitrate.[23] Stimulatory volatiles and phytohormones released by beneficial bacteria are immediately available for root absorption. Most telling of all, the lysis of mycorrhizal hyphae (see "Hyphal Lysis," page 25) — including the arbuscular structure right within the root cell itself — yields up lipids and proteins directly to plants in that everlasting flow of one life providing for the next. Plant nutrients do not all have to be reduced to simple ions in order to be assimilated as was once believed.

What I'm calling *green immune function* comes about for two reasons. First, metabolic energy reserves are a result of partially built nutrition delivered by robust soil biology. And, second, through contact with environmental reality. Occasional trials improve plant resilience just as much as human resilience.

Enter the beetles and plant bugs and snout-nosed weevils — the type that feed voraciously on foliage, lay eggs in developing fruitlets, bore into stems and trunks, and insert a proboscis into plump peaches. Japanese beetle, plum curculio, emerald ash borer, and marmorated stink bug, to name a few. All adults, all trouble. These insects have better-developed digestive systems and frankly don't give a damn about complete proteins. Disrupting genus priorities to feed and to reproduce on the part of higher-order insects requires that healthy plants conjure up antifeedant compounds and oviposition (egg-laying) deterrents in response.

Many well-documented insect antifeedants are in the triterpenoid chemical class. Alkaloid glycosides deter the larvae of the Colorado potato beetle in *Solanum* species, for one.[24] This sort of phase-four correlation is good stuff, but in truth our attention has been drawn more to rock star phytochemicals that certain plants produce, which can then be applied to crop plants. Several compounds in pure neem oil (pressed from the seeds of the *Azadirachta indica* tree) are hard to top in that regard. Salannin, nimbin, nimbidnin, and azadirachtin deter insect feeding. This synergistic thrust only works in *whole plant medicine* form (as opposed to commercial neem extracts narrowed down to a single reductionist constituent), so honor this herbal distinction. The insect-growth-regulating aspects of neem are totally about the azadirachtins.[25] Adult females are beyond this molting cycle impact, yet somehow sense "concern for future progeny" and thus choose to deposit their eggs elsewhere.

The beetle grouping is a tough nut to crack. Outstanding plant health certainly will tone down the roaring scream of *<name your own pest>*. Still, common sense dictates keeping sight of vulnerabilities in that pest's life cycle, supporting beneficial species through outrageous diversity, and making use of draw plants to trap the early comers before things get out of hand.

Plum curculio gets its dibs into many different fruits. Photo by Tracy Leskey, USDA Appalachian Fruit Research Station.

Similar advice applies for bacterial disease. Plant immune compounds are indeed capable of potent antibiotic clout. But here's the rub: Bacterial pathogens are opportunists. The initiation of any bacterial infection ties directly to openings into the vascular system of the plant. Forget the cuticle defense, forget enzyme inhibitors, and forget any notice to come up to speed. Every time a blossom opens, a pathway into the interior workings of that flowering plant lies exposed. Leaves torn by hail or driving rains are open to a bacterial thrust. Insects feed and plants bleed. Injury to bark tissues exposes cambium and thus intercellular sap flow. Disease organisms of the bacterial kind lie in waiting for precisely the right moment.

Bacterial wilt in cucumber, fire blight in pear, leaf streak on wheat, halo blight in bean, and bacterial canker on stone fruit can quickly get the show on the road. All such plant invaders (once established) have an overwintering mechanism that needs to be addressed. Abetting a healthier metabolism by way of trace mineral cofactors is good. Yet when push comes to shove, nothing beats an impervious crowd.

Dynamic Understanding

Now's as good a time as any to tackle those purist tendencies that come up whenever implementing one's chosen values. A teacher comes along with profound ideas about health, about farming, about the cosmos, about bliss. A recommended course of action follows from there. The underpinnings of healthy plant metabolism certainly suggest we can have it all. That somehow, praise be, enhanced mineral nutrition and honoring microbes forevermore will end the threat of pests and disease appearing on even a distant horizon. That somehow we ourselves need never get sick or, God forbid, experience cancer if we simply do morning yoga and drink nettle tea.

The truth lies more along the lines of *mostly*. And furthermore this is okay. Plants that encounter the full diversity of experience on this earth are more engaged phytochemically. People who breathe and eat and make love are toughened by those bouts of inflammation along the way. Immune function becomes all the stronger because of the occasional bump in the road.

Yet worse can happen. Nasties arrive from far shores to devastate fall raspberries. A perfect storm sets blight amok in the orchard. Influenza finds fault with the plans of even the fittest. Do we now continue to trust deeply held principles? Should we keep those original intentions intact?

The farsighted will tell you *mostly*. Being flexible to the situation at hand means you may need to instigate measures beyond everyday aims at times. A decade ago I stopped using sulfur to limit certain diseases in my orchard. And thus "holistic orcharding" came to be. Yet I still ponder this decision when likely circumstance (that stretch of fine spring weather followed by extreme rain) might suggest reaching for allopathic backup. I'll even make such a recommendation wholeheartedly to others in transitional situations. Contemplating change is a shake-your-bones proposition. Yet as nuance shifts and intuition deepens, it may indeed be prudent to temporarily adjust course when conditions warrant. Utilizing methods from complementary approaches is nothing like embracing the far side of the extreme.

A world without some flea beetles would be a big mistake. We need challenges to stay strong. Just like we need the flexibility to create localized answers based on dynamic understanding. The best teachers merely give inklings of how we might dance the dance.

Time once again to expand our sphere of consciousness. So far we have focused on biological reserves belowground. Let me introduce to you the one and only Arboreal Shield.[26] All plant surfaces are intended to be colonized by microbes. This part of environmental reality actually works in favor of plants. Only, stuff happens when you live in the great outdoors. From temperature extremes, drought, acid rain, and UV radiation to food resources running low. Throw in agrochemicals and you have the makings of a serious population crash. Arboreal colonization on plant surfaces can be diminished more than twentyfold by these factors. The other side of this story line is where things get interesting. Arboreal colonization greater than 70 percent denies both fungal and bacterial pathogens entry.[27] Keeping up *competitive colonization* with aerated compost tea, effective microbes, or indigenous brews is a core tenet of holistic crop management. This complementary "biological reserve strategy" is the best bet going for slamming doors on bacterial opportunists.

We move onward knowing that natural plant defenses are primed first and foremost by a diverse soil biology rich in mycorrhizal connection with a little help from some friends above.

Natural Plant Defenses

Plants have an immune system, of sorts. Rather than lymphocytes and natural killer cells, however, plant defenses are based entirely on resistance metabolites and essential oils. Here we turn the focus within to show how green immune function actually works and in what manner we as growers can encourage deeper metabolic riffs.

Two major tributaries feed the reservoir of multilayered plant immunity. The first branch results from direct exposure to pathogens and pests . . . the takeaway is (again) that plants are toughened by real-life experience. Molecular signaling emanating from infection sites initiates *systemic acquired resistance* throughout the affected plant. Volatiles

dispersed into the air forewarn neighboring plants that the time to deal with reality is now.

The second branch of immune response is triggered by biological and nutritional elicitors, leading to a heightened state of readiness at critical moments. The resulting array of defensive compounds (that prove so useful in herbal medicine as well) launched my interest in further working with *induced systemic resistance* for the plants themselves. Holistic foliar applications are crafted to stimulate an immunity cascade. What really catches my attention is how powerfully root-associated microbes kick this phytochemical response up another notch. Those fungi and bacteria that enter into the cellular space of roots indirectly activate pathways, which increase levels of resistance metabolites throughout the host plant before problems appear.

The evolution across biological kingdoms that points relentlessly to cooperation and support networks as the way to proceed in life is a lesson our species needs to take to heart. Going deep with this aspect of plant science will only scratch the surface of plant wisdom. The ecologist Frank Egler said exactly what I feel in contemplating all this: "Nature is not more complicated than we think — Nature is more complicated than we *can* think."

SYSTEMIC ACQUIRED RESISTANCE

The presence of a stressor launches the resistance route known just as readily by the acronym SAR. This trigger can be an insect bite, a fungal spore looking to get in its dibs, a parasitic nematode challenge, or even overgrazing by sheep and cattle.[28] The defensive compounds produced through systemic acquired resistance include more than 350 known substances, referred to collectively as *phytoanticipins*. These metabolites are called forth "in anticipation" of a pending challenge nearby.

The plant's response to whatever the incursion might be flows along these lines. An oxidative burst sets the salicylic acid pathway in motion. (Willow bark is rich in salicylic acid, and the reason why a chap named Bayer, in chewing on a willow twig for headache relief, ran with the idea of synthesizing aspirin.) This particular signaling pathway distinguishes systemic acquired resistance from other

SALICYLIC ACID RELEASE

INSECT INCURSION

PHYTOANTICIPIN BUILD-UP

Systemic acquired resistance (SAR) resulting from direct intrusion invokes the salicylic acid defense pathway.

defense mechanisms. One immediate result is the release of methyl salicylate as a signaling enzyme via both plant sap and as a fragrant scent into the air. This in turn prompts the buildup of salicylic acid levels in unaffected leaves as well as telling potential herbivores that they have been discovered.

The insect ramifications here are pretty cool. According to Ian Baldwin, director of the Max Planck Institute for Chemical Ecology in Jena, Germany, moths avoid laying eggs on plants that are giving off these volatile signals. The impregnated females either choose to avoid competition for their progeny or simply don't want to put eggs on a plant that is going to attract predators.[29] A single bite to one leaf is enough to deliver the message throughout the neighborhood.

Pathogens can adjust to salicylic martial arts more so than insects. The race is on once a disease spore lands on the surface of a leaf and sends out its own enzymatic announcement that the kid is back in town. Hyphal enzymes can suppress production of the phytoanticipins resulting from increasing amounts of salicylic acid in plant tissues. Fungi readily evolve to better tolerate phytochemical barriers. Phytoanticipins can be detoxified by means of counter enzymes. And sometimes disease fungi secrete virulence effector proteins to avoid being noticed in the first place.

Just as impressive, plants evolve in response to these strategies. Pathogen-associated molecular patterns (known as PAMPs) are the basis by which plant cells perceive an incursion. These are generally molecules essential to the pathogen that cannot be modified without significant loss of viability. Chitin in the fungal cell wall and the flagellin protein in bacteria are dead giveaways. Systemic acquired resistance will be a very active game board for a long time to come.

Yet plants can up the ante even more.

INDUCED SYSTEMIC RESISTANCE

The adaptive aspects of green immune function rely on entirely different signaling pathways. Both jasmonic acid and ethylene are integral to the "launching of a thousand ships" to further the cause of plant defenses.[30] Only these thousands of compounds as unveiled by induced systemic resistance (ISR) shall be known as *phytoalexins*. This major grouping of resistance metabolites accumulate in plants after exposure to benign microorganisms, foliar nutrients, and even physical contact. The formation of a single phytoalexin compound involves as many as twenty steps from the moment of perception by the plant. Once the signal has been received, genes within the plant are activated, which induce enzymes to begin the phytoalexin cascade. Methyl jasmonate or ethylene (both volatile hormones) will be released outboard from tissue, thereby invoking neighboring plants to respond in kind. Cell wall strengtheners, antioxidants, and pathogenesis-related proteins can be produced as well, depending on the stimulation source and plant in question.

The subtleties of a living soil system are many. Our latest collective discovery adds impressive weight to the importance of biological connection. It all began with the recognition that rhizobacteria had something more up their sleeve than "mere" nitrogen fixation. Turns out these root-colonizing microbes also stimulate systemic resistance. *Pseudomonas*, *Bacillus*, and *Trichoderma* species follow from there on the basis of intercellular presence in root systems. But the biggest honcho of all? Arbuscular mycorrhizal fungi.[31] Penetration of the root cell itself and the subsequent formation of arbuscules sends the jasmonic acid pathway into a tizzy. For the good of phenolic synthesis . . . and perhaps even world peace. The chemistry attributes may be many, but the fundamental thing to note is how mycorrhizal symbiosis delivers not only nutrients to host plants but immunity itself. *Now* we're really talking.

The term *elicitor* originally referred to applied compounds that stimulate phytoalexin production in plants. That's been broadened since in recognition of biologically induced systemic resistance.[32]

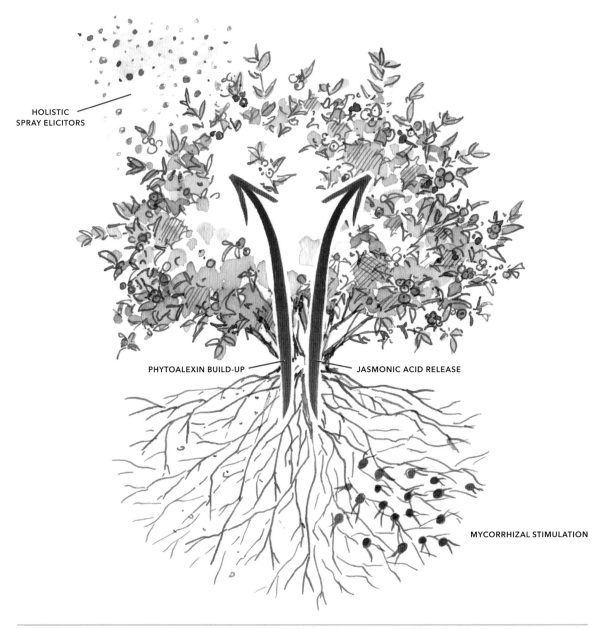

Induced systemic resistance (ISR) resulting from a nutrient or microbial elicitor invokes the jasmonic acid defense pathway.

In similar ways, arboreal microbes can "touch the right nerve" on the leaf surface and induce an immune response. Knowing this, let's flip the perspective by asking what we can do by way of foliar elicitors to call upon a heightened state of readiness when crop plants face foreseeable challenges.

Pathogens do not appear willy-nilly at any point in the growing season. Just as insects have cycles attuned to lengthening days and the warming of the sun, different fungi and bacterial disease organisms strike when climatic conditions warrant and prospects for success rate high. Young plant tissues tend to be more susceptible than time-toughened leaves further back on a shoot. The target area grows in sync with spore maturation as the plant crown or tree canopy fills out. Rots typically wait to make a final move until sugar levels rise as fruits start to ripen. Depending on the crop and the influence of the season — very wet being the worst, generally speaking — there are "windows" in every growing season when particular crops benefit from a deliberate boost to core defenses.

The terpenes, phenols, and alkaloids found in herbal remedies are a natural first choice for eliciting an enhanced response. *Like cures like*, preach the practitioners of homeopathy. That sort of thinking works in an energetic fashion (given the minute doses involved), whereas plant preparations applied to crop surfaces stimulate further production of comparable immune compounds. Resveratrol extracted from the roots of knotweed, for instance, isn't by any means the phytoalexin compound being induced in the target plant, yet this stilbenoid can generate equally desirable results.[33]

High up on the inducing elicitor list are compost teas and effective microbes. This extends the mission of the competitive crowd into the arena of in-house immunity. Job well done, guys!

Ionic minerals (such as Sea-Crop) contain eighty-nine periodic elements in balanced proportion, sourced from the living ocean. Seaweed extracts are good elicitors, in part because of the cytokinin, gibberellin, and auxin hormones in kelp. Purchased humus extracts (humic acid and fulvic acid) top off the list of holistic possibilities.

Now, it's time to get cooking, and only one rule applies: Activating multiple mechanisms with an assortment of foliar inducers is key. We can use the holistic core recipe (see "Fatty Acids in Garden Sprays," page 123) as a starting point. Typically, the effectiveness of foliar application to induce systemic resistance in the field stretches as long as ten days and up to as much as fourteen days in greenhouse trials. Fine-tuning an ISR approach requires giving plants a minimum of twenty-four hours' notice before a serious wetting event. In other words, foliar applications need to be made a day or two ahead of a fungal infection period or bacterial thrust to allow time for induced pathways to fully deploy. Prolonged lack of sunshine radically alters phytochemical oomph, no matter how you time it. As always, trace mineral nutrition is necessary for plants to synthesize the full range of phytoalexin compounds.

Essential oils have a special niche when it comes to plant defenses. These aromatic hydrocarbons primarily function as insect toxins, but also protect against fungal or bacterial attack. The mint family produces large quantities of menthol and menthone, which are produced and stored on the glandular hairs (trichomes) found on stems and leaves. Mints are rarely bothered by pests. Pyrethrins are monoterpenoid esters produced by chrysanthemum plants that act as insect neurotoxins.[34] Peel any citrus fruit and watch the fine mist display of limonene that results.[35] One of the essential oils made in the flowering cone of the hops plant is humulene . . . use of hop resins against

Infective Vernacular

Scientifically speaking, mycorrhizal fungi "infect" the roots of plants.

This word describes the reality of one organism entering into the space of another, often used in the context of microbes invading our own body tissues. And so disease gets passed along from one to the next.

Common interpretation of the word *germ* settles the matter that infectious diseases are caused by microorganisms. One of the earliest references to germ theory appears in *On Agriculture* by Marcus Terentius Varro (published in 36 BC) who warns about locating a homestead in the proximity of swamps:

> *. . . because there are bred certain minute creatures that cannot be seen by the eyes, which float in the air and enter the body through the mouth and nose and there cause serious diseases.*

Bacteria and viruses lead the way here, yet the kingdom Fungi often get tainted by this same linguistic broad brush. Mold growth on peaches, loss of a barley crop to powdery mildew, tree mortality due to white pine blister rust — all are the result of pathogenic fungi finding a stressed niche to call home.

What's fun about human understanding are those essential shifts in perception made now and again. Seeing the same old same old with new eyes opens the doors to innovative approaches across the board. Consider our microbiome, the community of a hundred trillion microbes living in the human body. These allies protect us from pathogens, help with digestion, synthesize certain vitamins, and regulate our immune system. Similarly, healthy plants have equally important microbial connections, from the root zone all the way up to those leaves waving in the bright sunlight. Fermented foods and biological farming follow from there.

Shall we reconsider the language with which this all began? To *infect* can just as well mean to influence, to inspire, to enthuse. Those microbes misconstrued as *germs* are the origin of life, the seeds of our continuance, the spark urging on immune function.

Mycorrhizal fungi "infect" the roots of plants . . . thereby triggering induced systemic resistance from below. Praise be.

bacterial canker on stone fruits works because these terpene compounds prove to be antibacterial.

Strengthening the immune response of plants by applying foliar elicitors increases the amounts of resistance metabolites available. Take this thinking one step further. Could it be that foods and healing herbs grown "Nature's Way" turn out to be all the more potent for us as well?

Plant Metabolites and Human Health

Let's talk about phytonutrients that make our lives healthier. The deep reservoir of immune compounds produced *by plants for plants* finds dual purpose in helping our bodies ward off chronic disease and dial back those inevitable downtimes. Herbalists use plant medicines to soothe tissues, reduce inflammation, strengthen the heart, cleanse the liver, quell depression, heal festering sores, set bones, and boost immune response to this extraordinary world. Nutritional support for body systems in turn enables self-healing. That seemingly endless list of conditions and symptoms and general lack of pep faced by our species today reflects the enormous shortcomings of industrial agriculture as well as a serious disconnect with our gut microbiome. The simple act of taking bitters stimulates proper digestion. Get elimination back on track and . . . but wait. This isn't an herb book, is it? Apparently I digress.

The very best herbal advice falls right in line with the old adage to "let thy food be thy medicine." Nutritious food is medicine of the right sort, which builds health day after day. Back to thoughts of vibrancy, vitality, and *joie de vivre*! We need those resistance metabolites found in plants, which result primarily from a robust soil biology. That the plant realm and the microorganism realm and the human realm correlate should not surprise anyone. We come from the same source, the same ways of being, the same fundamental chemistry.

Plant constituents can be classified on the basis of molecular structure, composition (containing nitrogen or not, for instance), solubility in various solvents, or the pathway by which compounds are synthesized.[36] We'll continue to keep this simple and focus on three main groups. Distinguishing terpenes, phenols, and alkaloids by listing a few of the medicinal properties found within each class just might drive home the point as to how "unsecondary" these metabolites prove to be.

Terpenes occur in all plants and include over twenty-two thousand compounds containing carbon atoms in multiples of five. Compound that molecular math and you have the active constituents in plant resins and essential oils. The broad view of terpenes and terpenoids includes preemptive bearing on cancer, antimicrobial and anti-inflammatory activity, and even putting the kibosh to internal parasites. Essential oils have exceptional skin permeability and so are applied topically, mostly. Tea tree oil, for example, can be used for acne, skin fungus, cold sores, and other topical infections. An extract of feverfew helps with arthritis, migraine headache, toothache, and menstrual pain relief, in part due to parthenolide, a sesquiterpene. Medical marijuana users are actually appreciating terpeno-phenolic compounds in the form of cannabinoids. Terpenes give cannabis its scent and flavor . . . and, once coupled to a phenol, help provide abundant levels of mood-altering THC.

Phenols are derived from simple sugars and contain benzene rings as part of their molecular structure. The wide range of phenolic compounds includes flavonoids, anthocyanins, tannins, and coumarins. Flavonoids act as pigments, imparting color, often yellow or white, to flowers and fruits. Such polyphenols are antioxidant, which is why a tea made from hawthorn blossoms (with basal leaves intact) is an excellent circulatory tonic for the heart. Anthocyanins provide the darker-colored pigments found in berries. These free-radical scavengers tone down oxidative stress, thereby helping to prevent atherosclerosis, inflammatory conditions, and certain cancers. All plants produce tannins to a greater or lesser degree. Astringent polyphenols like these help contract body tissues and dry up excessive watery secretions. Pleasant-smelling coumarins found in plants such as licorice and meadowsweet help support connective tissue, modulate inflammation, thin the blood, and reduce lymphatic swelling.

Alkaloids are a large class of bitter-tasting nitrogenous compounds that are found in some vascular plants. Many of these compounds are derived from amino acids and have powerful effects on animal physiology. Think cocaine, morphine, and nicotine. Notoriety aside, the use of quinine from the bark of the cinchona tree has been used to treat malaria since at least 1632. Piperine gives black pepper its pungency, and when cooked up with turmeric and ghee in Indian food, oh man, synergistic medicine never tasted so fine.[37] Capsaicin in chili peppers produces the "internal fire" needed to prevent sinusitis and relieve congestion. The best-known alkaloid of all, caffeine, is responsible for the jolt in every cup of coffee you ever drank.

Food as medicine comes with a caveat. How we grow our fruits, vegetables, grains, and herbs matters. The full range of healing constituents comes about in living soil systems from plants confronted with environmental reality.[38] See chew holes on the edge of the arugula? Or a spot on that apple? Wahoo! Give thanks to a sophisticated defense strategy on the part of plants, which can only mean more stress molecules for us.

CHAPTER 3

Underground Economy

All that phytochemistry in the previous chapter springs from the uptake of nutrients in the soil below. Time to examine the *fungus-root* relationship in the context of the countless other organisms that make up the soil food web. Mycorrhizae are involved in a "carbon currency" with bacteria in making nutrient flow happen for green growth aboveground.

"Heaven is under our feet as well as over our heads," wrote Henry David Thoreau in *Walden*, back in 1854. Hardly a single facet of what we are about to explore was understood then . . . yet this expression of veneration continues to be entirely apt as we unveil the workings of the underground economy. Others have said that the soil biology digests and provides the nutrients needed by plants, much like the microbes in the rumen of a cow. Nor are we humans without our own essential gut microbiota. Plants developed roots in the course of time as a similar interface for teamwork.

Root exudates rich in carbohydrates and proteins draw many interested parties into the rhizosphere. A range of numbers describes the intensity at play here. Plants direct 20 percent of photosynthesis sugars to mycorrhizal trade. Plants send 40 percent of all photosynthesis production to the rhizosphere to stimulate microbial interaction. Plants send nearly two-thirds of carbon synthesis down to their roots once green metabolism becomes fully engaged. All of this involves some guesswork, but the point is well taken: Plants have carbon currency in hand to attract proven allies.

This availability of carbon sugars results in as much as a hundredfold increase in microbial density in the immediate root zone and a microbial composition that is far more diverse than in the surrounding bulk soil. Any trading partner worth his or her salt knows how to swing deals. Plants are no different in this regard. The composition of the root biome can be modulated by making compounds available to either feed or suppress the interested parties at hand. An additional charge of carbon comes in the form of root cap cells that protect delicate root growing tips (meristems). New cell growth pushes the root forward, causing older cells on the periphery of the root cap to rupture and fall away. This root mucilage becomes fair game for soil microorganisms and thus is why the greatest concentrations of bacteria are found alongside root tips.

The bacteria and fungi that consume carbon-rich exudates in turn are eaten by bigger microbes, principally nematodes and protozoa. Every microbe enters the food chain in due time, providing sustenance for the next round of organisms. Minerals taken in (assimilation) become minerals released (mineralization). Roots are awash in this flood of nutrients by virtue of attracting all this activity into the rhizosphere in the first place.

Roots come into contact with nutrients by one of three means. *Direct interception* is responsible for an appreciable amount of calcium uptake in certain soils, followed by magnesium, zinc, and manganese. This set of nutrients is made available by the action of organic acids, chelates, and other compounds produced by roots as well as bacteria drawn to the scene. *Mass flow* occurs when nutrients are transported to the surface of roots by the movement of water in the soil. Such capillary action shuts down under drought conditions. Most of the nitrogen, calcium, magnesium, sulfur, copper, boron, manganese, and molybdenum move to the root by mass flow. *Diffusion* delivers a particular nutrient by means of a concentration gradient. This tendency to seek equal distribution keys to the availability of the select nutrient in the surrounding soil, especially in the case of phosphorus, potassium, and iron. Diffusion is a relatively slow process when compared with water movement toward the root.

Root-soil contact is determined by root length, root branching, and root hairs. Root hairs are located just behind the root tip and have a relatively short life span of a few days to a few weeks. Actively growing feeder roots are necessary to continually renew these prime locations for nutrient uptake.

Still, plant roots can only "shop locally" for so long, especially when it comes to less mobile nutrients. Dig up a mass of roots, and what appears to be a considerable surface area for absorption only reaches so far. Take this to the macro level and even a fibrous root system occupies at most 3 percent of the total soil volume. Left to their own devices, roots can access 1 to perhaps 2 cubic centimeters beyond the actual root epidermis to directly uptake mineralized nutrients. Not unlike being given a short straw to drink a root beer float from a tall frosty mug on a hot summer day. A *nutrient depletion zone* develops where no amount of urging by the hungriest of root hairs can lap up what's no longer there.

Limits being limits, roots are quite willing to work with fungi offering a vast mycelium of "flexible straws" to reach a broader array of nutrients. Fungal partners can reach as much as a hundred times more available nutrients due to the *surface volume extension* of mycorrhizal hyphae. Spatially speaking, plants can access 12 to 15 cubic centimeters of soil around individual roots through this mutual association. The rhizosphere in essence enlarges to become what's called the *mycorrhizosphere*.

And thus things would stand were this exclusively about one mycelium, one plant. Successful regional economies, on the other hand, involve resources being shuttled back and forth through networks according to supply and demand. Many mycelia, many plants . . . starting with even more distributors bringing even more nutrients into play.

Dancing in the Street

The soil food web involves trillions upon trillions of critters consuming first organic matter, then each other, and releasing nutrients in the process.

UNDERGROUND ECONOMY

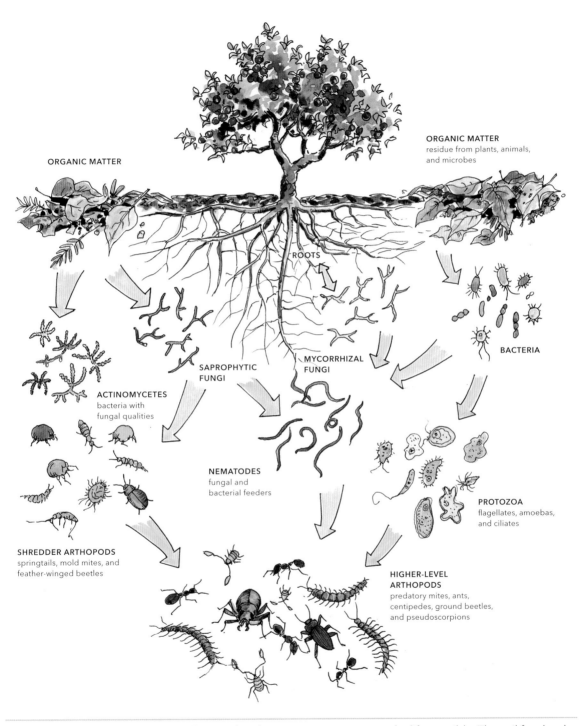

ORGANIC MATTER

ORGANIC MATTER
residue from plants, animals,
and microbes

ROOTS

SAPROPHYTIC FUNGI

MYCORRHIZAL FUNGI

BACTERIA

ACTINOMYCETES
bacteria with
fungal qualities

NEMATODES
fungal and
bacterial feeders

PROTOZOA
flagellates, amoebas,
and ciliates

SHREDDER ARTHOPODS
springtails, mold mites, and
feather-winged beetles

HIGHER-LEVEL ARTHOPODS
predatory mites, ants,
centipedes, ground beetles,
and pseudoscorpions

Interdependent and interconnected networks of organisms interact to make life possible. The soil food web encompasses the microbes and arthropods that ultimately provide balanced mineral nutrition for plants and thus promote healthy plant metabolism. Illustration by Elayne Sears.

Quite the social scene, that's for sure. Around 80 to 90 percent of plant nutrient acquisition is microbially mediated at one point or another. So just who are the other players on what I like to call the fungal team?

This thriving community of organisms in the soil can be sketched out along nutritive lines defined by *who eats what*, and *what eats it*. Chances are you've seen this organized according to the different trophic levels of what's essentially a big feeding frenzy. Things kick off with the main course — organic matter created by plants through photosynthesis. The second trophic level consists of the decomposers, mutualists, pathogens, and parasites. Mycorrhizal fungi join in with other fungal kin, actinomycetes, and countless bacteria to begin the conversion of organic matter back toward life. The third trophic level finds the shredders, predators, and grazers at work. Fungal-feeding mites join with fungal-feeding nematodes and bacterial-feeding protozoa (flagellates, amoebas, and ciliates) and bacterial-feeding nematodes to ramp nutrient availability into high gear. Higher-level predators that we can actually see take things from there, including roly-polys, spiders, millipedes, centipedes, earthworms, and ground beetles.

The process of decomposition is certainly as important as the process of photosynthesis. One force builds up what the other breaks down, thereby creating our interconnected and interdependent world. A teaspoon of soil may hold ten thousand to as many as fifty thousand different types of bacteria, along with thousands more species of fungi and protozoa, nematodes and mites — perhaps even one incredibly stressed ant. Obviously, mycorrhizal fungi tap into the work of others in snagging nutrients for plants. Let's list several direct tie-ins between symbiotic fungi and the other decomposers in the soil food web.

Bacteria break down plant residues differently than fungi and have distinctive roles in the recycling of nutrients. Communities of "helper bacteria" congregate along mycorrhizal hyphae, where they can transfer elemental nutrients for use by plants in exchange for a food source. This secondary trade in carbon gets a good return, for these assorted bacteria are involved in supplying eleven of the fourteen nutrients required by plants. Mining rock (mineral weathering) is a joint effort between "bacteria bore species" and ectomycorrhizal hyphae . . . just as endomycorrhizal hyphae are totally down on phosphate-solubilizing bacteria. Meanwhile, nearer the root side of the mycorrhizosphere, where microbe numbers soar, other bacteria tend to numerous matters, including the root-colonizing rhizobacteria. Mutualism adds a third leg in this case, with the phosphorus delivered by mycorrhizal fungi to the root system enabling greater fixation of nitrogen by the rhizobacteria.[1]

Bacterial membranes consist of energy-rich phospholipids that degrade easily and quickly in the jaws of ciliates and the like.[2] The absorptive surface of extraradical hyphae taps into this action in a big way to deliver nutrition to plants in more complex forms. Bioavailability of lipids (as fatty acid components) stimulates mycelial growth as well, thereby enhancing mycorrhizal formation.

More bacterial connections with mycorrhizae can be found by turning inward. Bacteria tuck away within the mantle and nutrient-exchange structures of ectomycorrhizal fungi. Similarly, arbuscular-type hyphae have been shown to host intracellular colonization by bacteria. This may simply represent an opportunistic colonization of damaged hyphae, or it may constitute endosymbiotic associations. Bacteria can even be found in the cytoplasm of AM fungal spores . . . the scale of which goes beyond small (see "Inoculum Nuance," page 89), as we'll learn later. Keeping a relative sense of place in all this gets mind-boggling, without question.

Actinomycetes straddle the species line between bacteria and fungi. Actinomycetes were long considered to be fungi based on morphological features such as hyphae and spore formation. Analysis of DNA suggested otherwise, however, and thus actinomycetes are now recognized as filamentous bacteria.

These aerobic organisms contribute significantly to the decomposition of cellulose. Plant fiber has a high carbon content and a correspondingly high carbon-to-nitrogen ratio, making it an ideal food resource for both decomposer and ectomycorrhizal fungi. Many strains have the ability to solubilize inorganic phosphorus or make available organic phosphorus sources. Actinomycetes perform like helper bacteria as well, by improving AM mycelia outreach. Conversely, these microbes can decompose chitin and thus play an important role in nutrient release by extraradical hyphal lysis.

Soil fungi are best grouped by function. But be forewarned: Species lines cross over with apparent ease when it comes to parasitism, just as occurs with out-and-out mycorrhizal types on the mutualistic side. *Live Free or Adapt* is the fungal maxim that keeps things interesting.[3] The second trophic level discussed at the start of this section provides the very framework required for this task.

Decomposers are the *saprotrophic fungi* that break down organic matter of all kinds, including wood and every other type of plant material.

Mutualists are the *mycorrhizal fungi* that form this cool symbiotic relationship with plant roots. You'd sure better know a bit about these blokes by now!

Pathogens are disease-causing fungi, including the well-known *Verticillium*, *Phytophthora*, *Rhizoctonia*, and *Pythium* species. These penetrate the roots and decompose living tissue, leading to weakened or dead plants.

Parasites are *saprophytic fungi* that manifest disease symptoms if conditions warrant, while also breaking down organic matter.

This last category, the parasites, is in desperate need of some clarity. Let me tell you why. Everyone — and I indeed mean both *him* and *her* and myself included — has spoken for the longest time about "saprophytic fungi" as the ones that decompose organic matter. Yet the botanical root of this word clearly lies with *phyte*, which means 'plant' as derived from the ancient Greek. As in a living, growing, green plant. It's the prefix *sapr-* that steers things in a 'putrid' direction. Think dead plant residue and allow for a slightly sweeter smell, perhaps. Those of you champing at the bit to complete this word play need to know that *troph* meant 'feed' to those same long-gone Greeks. Clearly that gives the edge to saprotrophic fungi as the ones feeding on rotting organic matter. But what to make of the alleged "putrid plant" fungi?

Turns out that we have need of just such a word to describe fungi with a dual nature. There are numerous fungi that feed on dead organic matter but also take a parasitic bite from living plants on occasion. These are the true saprophytes.[4] Quite a few live inconspicuously in the root, inflicting little damage on the host, seemingly more saprotrophic than not. *Armillaria* species, however, can attack entire root systems of trees, causing the whole plant to die, but then carry on to break down

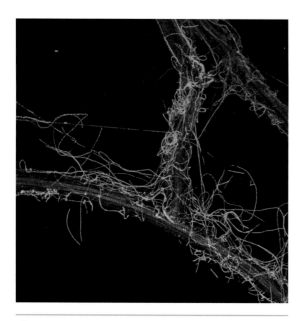

Fluorescent microscopy image of a root of *Arabidopsis thaliana* surrounded by hyphae of *Colletotrichum tofieldiae*. This endophytic fungus has been granted symbiotic access by the cress in exchange for phosphorus. Photo by Ryohei Thomas Nakano, Max Planck Institute.

the resulting organic matter. That phase seems harmless enough – consuming rotting wood – but then under certain conditions these same fungi spread onward to uninfected roots by means of rhizomorphs.[5] The term *facultative saprophyte* accommodated such situations when saprophytic everything carried the day.

The *endophytic fungi* take adaptability to yet another level. These species work within plant tissues, often acting benignly, yet capable of crossing lines into mutualism or parasitism. Things get especially interesting in the case of *Colletotrichum tofieldiae*, only recently caught in the act of stretching apparent plant boundaries. Researchers discovered that this fungus was bringing phosphorus to thale cress on the central plateau of Spain.[6] Sounds like a mycorrhizal kind of thing to do . . .

except *Arabidopsis thaliana* is a nonmycorrhizal plant anywhere else. The cress altered its immune response to this fungus in order to trade for phosphorus in low-phosphorus soils. Synthesis of mustard oil glycosides by the cress would otherwise prevent this fungal association entirely. According to Paul Schulze-Lefert, director of the Max Planck Institute for Plant Breeding Research in Cologne, "It's an elegant solution that extends the role of the immune system to ensure an external supply of nutrients under malnutrition conditions. This has not been previously observed in the plant kingdom."

We can even throw some revered mutualism into the pot if we really want to have fun. The common morel and related species are considered to be mycorrhizal (see page 185 in chapter 7) yet are noted for mysteriously appearing in their fruiting phase around burn sites and dead elm stumps only after the tree hosts have passed on. The game is afoot for the "mushroom of our desire" . . . but let's save that one until we go shrooming.

Fungi are the most abundant group of soil microorganisms on a biomass basis. Some seventy thousand species have been described, but at least twenty times that number is estimated to exist worldwide. Fungi have evolved extracellular digestive enzymes (cellulases, laccases, betaglucanases, mannanases, and so on) to digest plant material. Hyphae in turn are consumed, thereby making nutrients available to higher trophic levels. Mycorrhizal fungi absorb a portion of these mineralized nutrients as the world turns. What might be called the "universal mycelium" involves a lot of teamwork.

The dance continues as root exudates, fungal hyphae, and the secretions of bacteria bind small soil particles and organic matter together to improve soil tilth. This provides a better soil habitat that attracts more critters, which further increases the amount of nutrient cycling. Castings from earthworms and fecal pellets from macroarthropods increase the number of larger-sized soil aggregates, allowing for improved water infiltration and aeration. Plant roots extend further into the ever-richer earth. Tunneling by soil animals mixes organic matter particles deeper into the soil, thereby increasing the water-holding capacity of the soil yet again. All is good.

Mycorrhizal Networks

A healthy ecosystem features outrageous diversity. Each one of those plants is connected to others of its kind and essentially every other plant in sight by mycorrhizal hyphae. One fungal network with affinity for certain plants in a given ecosystem invariably includes "passage plants" with multiple affiliations with other mycorrhizal networks. These shared vascular systems in turn become transfer sites where different species of AM fungi can pass along nutrient deals to each other. Ultimately we'd be right to consider *all as one* as different fungal networks form multiple affiliations with multiple plants. Every base will be covered in the **common mycorrhizal network** that develops when healthy soil function is kept to the fore.

Let's spell this out in greater detail from the perspective of mycorrhizae. The trees in a temperate forest are interconnected by multiple ectomycorrhizal fungal networks. This holds true for a broad mix of species, from oak to pine to birch and onward. Meanwhile, understory shrubs, grasses, and herbs are interconnected by multiple arbuscular fungal networks. Many of those AM fungi tie in as well with select trees, particularly softer hardwoods such as maple and poplar, which act to "bridge" the two fungal realms together. Some communities even feature a third and a fourth *fungal canopy* formed between ericoid and orchid mycorrhizal plants. Add fungal adaptability to this mix (EM fungi assuming ericoid traits, for instance) along with the ready acceptance by plants of nonspecific overtures . . . and we're looking at a fungal map where no plant is left out of the loop. Just as impressive are the seemingly limitless ways to shuttle from one point to the next. Can you imagine landscape architects taking on a design project this involved?[7]

The complexity and composition of mycorrhizal networks change over time as plants germinate, grow, compete, respond to seasons, are consumed, and die. Similarly, nutrient resources are in flux as needs are met, microbe populations shift, depletion zones develop, and hyphae access untapped veins of mineral wealth. Movement of nutrients from plant to plant in ecologically meaningful amounts has been documented again and again. Nor should we forget that new avenues are a blink of the eye away if need demands.

A single fungal network develops spontaneously as multiple spores germinate and hyphae search to discover roots. Anastomosis encounters of a similar kind (see "Fungal Adaptability," page 14) allow hyphae to tap into one another, thereby spreading the circumference of absorption and assimilation. Such hyphal fusion makes possible the formation of a clonal colony that can eventually cover a significant area, from 1 hectare (2.5 acres) on up. The frequency of anastomosis formation between extraradical hyphae spreading from the root systems of different plants is the means by which plant interconnectedness becomes complete.[8]

Herein lies the shaping of ecosystems. Rapid root colonization determines plant species composition, in part, by securing the niche of selected species with nutrient support. Ectomycorrhizal colonization of seedlings by a mycorrhizal network

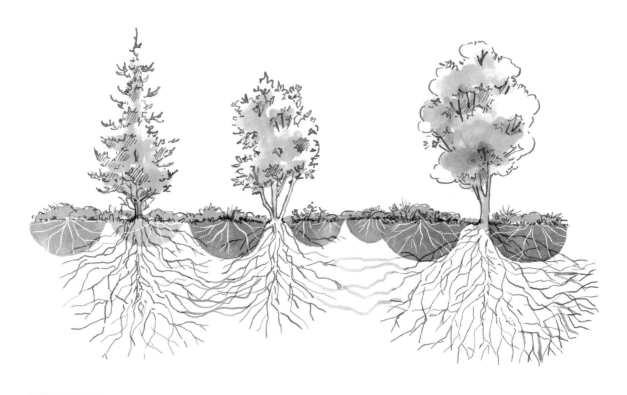

The common mycorrhizal network leaves no self-respecting plant out of the support loop. Passage plants, bridge trees, and fungal adaptation make possible nutrient exchange and messaging throughout a plant community.

of older trees, for instance, is critical in that first year of growth when the next generation seeks to get established on the forest floor. As the medium of soil structure, mycelial networks determine the flow of water and air, literally directing the pathways of root growth and even opening channels for the movement of soil animals. Mycorrhizae determine the metabolic processes of the soil by moderating the microbial community as a whole. Certain networks may even be responsible for facilitating the transport of allelopathic chemicals from especially insistent plants directly to the rhizosphere of competitors.

Let's look at a few practical applications of mycorrhizal networking in our own doings with plant ecosystems.

Planting a "cover crop cocktail" involves a mix of different plants that work together to improve soil. Legumes such as clover or field pea are included to fix nitrogen. Cereal grasses such as oats or millet run deep roots and thus bulk up organic matter belowground. Certain flowering herbs, such as dill, may be chosen to support beneficial insects. Good farmers think they have a handle on *who, what, and why* . . . but the full story line holds networking intrigue as well. Legume nitrogen gets passed along to the grass companions through fungal hyphae. A trade is made, in essence, as arbuscular fungi bring phosphorus into the network on behalf of the grasses. The cereal plants send a phytochemical signal via the fungi to absorb surplus phosphorus against the diffusion gradient, "knowing" this is the means to

Outrageous diversity that creates a beneficial insect haven above creates a resilient mycorrhizal network below in this young pear orchard. Photo by Linda Hoffman.

get more nitrogen. The rhizobacteria fixing that nitrogen on behalf of the legume are the actual recipients of the phosphorus, thereby saving procurement energy in order to fix more nitrogen.[9] The upshot here is that a monoculture of legumes would be far less efficient — and thus less productive — without the grass.

Out in the orchard, apple and pear trees enter the reproductive phase when fruit sets. Carbon trade is at full bore at that point, engaging the spring root flush (see "Pulsing the Spring Root Flush," page 149) to take up minerals galore for developing seed as well as embryonic flower cells for next year's crop. The fruit-set window takes thirty to forty days to complete, at which point feeder roots recede and shoot growth resumes. Carbon sugars are

then directed above and, accordingly, mycorrhizal activity slows down for the summertime. Terminal buds send the signal to prepare for winter by some point in August (ending the second phase of shoot growth), thereby initiating the fall root flush. Can you guess where "tree carbon" is going? Keeping the mycorrhizal network fully engaged through the early fall months for this important period of mineral uptake by the trees means that a majority of understory plants area-wide should ideally be in the vegetative stage as the harvest begins. Grasses focused on renewed growth are not setting seed. Correspondingly, understory mycorrhizae are not chiming in for minerals to the same degree as the fruit trees. Aisleways that have been grazed, mowed, or planted to cover crops in midsummer help keep "understory carbon" out of the trading loop at this juncture. Orchard management decisions are in truth about pulsing fungal incentives to achieve optimal production.

Forest succession has far more to do with mycorrhizal networking than any other factor. Fungal mycelia guard the forest's overall health by multidirectionally allocating nutrients. A young pine stand with alder counterparts plugged into the same ecto network can get the nitrogen it needs. Douglas fir seedlings and paper birch shuttle carbon back and forth to each other seasonally via shared hyphal pipelines. Paper birch takes care of the fir seedlings when these upstarts are shaded in summer. Come spring and fall, the Douglas fir returns the favor when the birches have no leaves. The role of "mother trees" in caring for their very own especially highlights how important fungi are in sustaining the next generation. Trees can genetically recognize offspring as kin through the mycorrhizal network (*boing*!) and subsequently direct nutrients to shaded seedlings until they're tall enough to reach the light.[10] Similarly, any remaining semblance of roots in a mixed stand that's been clear-cut will give its last breath to the cause of reestablishing new trees. Stump sprouts per se may not all go the distance, yet they attempt to rectify what humans have destroyed. Nature wants to keep space filled.

Innate Intelligence

When we try to pick out something by itself,
we find it hitched to everything else in the universe

— *JOHN MUIR*

The overall mycorrhizal network has been called the "wood wide web" in reference to our less complex global computer network. Keep it humble, human.[11] The first function of this superorganism is to move nutrients and water between plants of different species. Balanced nutrition results when fungi deliver what plants really need. Providing immune alerts community-wide allows plants time to prepare for imminent challenges. Fungi maintain bidirectional communication on all such matters pertaining to ecosystem resilience.

The acumen to promote system vitality becomes readily apparent in wilder places. Health abounds, whereas in cultivated landscapes the greenness somehow seems that little bit more contrived. Stinging nettle demonstrates this contrast perfectly. Volunteer seedlings that pop up out in the orchard or in abandoned pastures are always a far deeper shade of green than cultivated plants started in the greenhouse and then grown for medicinal herb production. The heart may be willing, but it still comes round to having friends in the right places. The enhanced vigor of natural landscapes reflects a fully functioning mycorrhizal network at work.

That this "fungus-plant intelligence" comes versed in phytochemistry and cellular polarity should not deter our recognition of an equally perceptive system. The human brain lends our species its own take on judgment and knowledge, but any outsider could quickly reveal that we, too, function basically by means of biochemistry and neurons.

Plants have evolved certain senses, including the five we consider our own. The ability to *smell* and thereby *taste* comes by way of volatile chemical signals in the air. Reacting to various wavelengths of light alludes to *sight* by means of photoreceptors. Just watch a flower track the path of the sun across the sky and then close when day ends. Plants *touch* when roots instinctively detour away from an impenetrable barrier or a vine deliberately grabs hold of a trellis wire. The ability of plants to *hear* is more of a stretch. Reports suggesting classical music abetting lush green growth come with an element of wishful thinking. Yet sound waves have been associated with yield increases in vineyards in Tuscany, just as the vibrations of gurgling water in a buried pipe attract roots.[12] The tips of plant roots are noted as well for being able to sense gravity, moisture, soil compaction, and mycorrhizal presence. Similarly, roots sniff out specific mineral sources while avoiding areas with excessive concentrations of carbon dioxide, salinity, and toxic metals.

Much is made of the fact that plants are not mobile and thus so unlike us. Oh, the assumptions we make! Our perspective as human individuals able to run amok skews any interpretation of what it might mean to be rooted in this earth. The plant community as a whole consists of interconnected beings making collective decisions that go beyond the one. Mycorrhizal networks provide speed and fluidity to each member of the community. This different form of motion — that of emergent cells exchanging signals and nutrients — brings things onto a more even keel. The common good can be found precisely because *living intelligence* is at the helm.

BALANCED NUTRITION

Plants get 95 percent of the elements they need from air and water. Carbon. Hydrogen. Oxygen. The other 5 percent comes from the soil. The whole shebang of mineral elements, including nitrogen in plant-friendly form, comes by way of microbes and direct absorption. Roots snag a smaller proportion of minerals directly in natural systems if the chance for mycorrhizal trade is on the table. All such decisions are economical, in a sense. Producing organic acids to dissolve rock requires more work on the plant's part than allotting "free" photosynthesized

carbon sugars for microbes to do the work. The common mycorrhizal network in essence provides a single nutrient-uptake system offering the best deals in town.

Many things influence transfer patterns. Photosynthetic rates determine the amount of ready carbon to trade. Growth rates are about keeping more of that carbon for plant requirements. Defoliation by pathogens or insects bumps the afflicted to the back of the line when it comes to the ready availability of carbon currency. Drought slows down metabolism . . . providing fungi with the incentive to bring moisture in order to keep photosynthesis engaged and thus carbon flowing. These affairs would seem to be on track to favor the strong, wouldn't you agree?

The *law of demand* has much to do with which plants get preference for the nutrients available from fungi. Put on your counterintuitive thinking caps, please. Fungi will pay less for carbon as the quantity available from a high-performing plant rises. This in turn affects the *law of supply*, as healthy plants will provide additional carbon for fewer minerals. What's happening here explains the altruistic nature of mycorrhizal networking. Hyphal pipelines move nutrients from areas where resources are high to areas where resources are low. Plants accumulating large nutrient reserves are weaned from gluttony (more or less) in consideration of seedlings and other plants struggling to gain resources. The underdogs get carbon from fungi that have more than enough. We might just call this "social democracy" at work if we wanted to get political about our applied biology.

Plant ecologists have had to rethink the concept of each plant being an individual competing with its neighbors for the same resources. The benefits to the community as a whole turn out to have value for the individual as well.

The Fungal Seer

Rudolf Steiner gave a series of agricultural lectures nearly a century ago that became the basis of biodynamic farming. His suggestions are not without controversy, as Steiner navigated his way primarily from a spiritual and mystical perspective. Biodynamic growers use specified herbs and minerals for compost preparations and field sprays, burying such in cow horns and animal organs for a fixed time in the making, often in accord with planetary phases. Steiner felt that these remedies mediated terrestrial and cosmic forces in the soil.

Let's take a few Steiner quotes head-on. The interpretations chosen appear straightforward; especially with the fungal context provided . . . just know there are appreciable challenges in translating individual words from the original German that embody multidimensional lines of thought. One only broaches Steiner with concerted effort!

Rudolf Steiner in 1911. Courtesy of the Rudolf Steiner Archive and e.Lib.

> *For many plants there is absolutely no hard and fast line between the life within the plant and the life of the surrounding soil in which it is living.*
>
> *LECTURE 4 (ADAMS TRANSLATION)*

Steiner had just been speaking about how the trunk structure of a tree would better be viewed as a reaching up of the soil life, that

Healthy plant metabolism requires access to a wide range of elements. The complexity of enzyme synthesis makes certain trace minerals all the more precious. Guess what you do if you're rich in carbon? You willingly splurge. And in this case the resulting cascade of phytochemical progression indeed trickles down and outward and back to all. Fungal networks with bidirectional streaming capability ensure distribution of nutrients more evenly. Trades are made on a case-by-case basis, yet there's plenty to go around thanks to the frenzied crowd to be found in every the plant aspect of a tree sprouts forth from this raised earth we call bark. The biological implications here proved profound for my own holistic orcharding. This statement further discerns a mycelial connection for most plants at a time when others thought primarily of fungi as parasitic.

Treat the soil of the earth as I have now described, and the plant will be prepared to draw things to itself from a wide circle. Your plant will then benefit not only by what is in the tilled field itself, whereon it grows, but also by that which is in the soil of the adjacent meadow, or of the neighboring wood or forest.

LECTURE 5 (ADAMS TRANSLATION)

Steiner's teachings this particular day of the course centered on the use of six herbal preparations to enliven compost. The flower of dandelion, for instance, is to be considered "a tremendous asset because it mediates between the fine homeopathic distribution of silicic acid in the cosmos." Right, that. Yet somehow biodynamic compost intentionally made with the compost preps — consisting of yarrow, chamomile, nettle, oak bark, dandelion, and valerian — somehow, this enlivened organic matter helps plants reach nutrients beyond discernible root reach.

The correct balance of woods, orchards, bushes, and meadows — with their natural growth of fungi — is so essential to good farming that your farm will really be more successful even if this means a slight reduction in your tillable acreage. There is no true economy in using so much of your land that all the things I've mentioned disappear.

LECTURE 7 (CREEGER/ GARDNER TRANSLATION)

Sounds like a plug for outrageous diversity, along with a first explicit recognition of the fungal element. Steiner was obviously alluding to the hyphal reach of mycorrhizal fungi at this point. Still, the year was 1924 . . . and human understanding of the world beneath our feet was just getting under way.

A core tenet of the Steiner philosophy holds that intimate discovery is available to every conscious individual. Each of us can access imagination, inspiration, and intuition in working with fungi and healthy plants. Many paths lead to the mycelium.

rhizosphere. Partially built nutrition in the form of bacterial metabolites and hyphal lysis notches up plant metabolism yet again.

That's a lot of words to say what Graham Phillips at BioOrganics says in relatively few words: "Mycorrhizal fungi offer a regulated uptake of nutrients at a good pace rather than an injection of conventional synthetics."

The flip side of balanced nutrition is imbalanced nutrition. Introducing a flood of soluble nutrients by way of chemical fertilization results in excess nitrates and a likely reversal of important fertility ratios that improve plant resistance to disease. Throw protein synthesis out of whack and there's a price to be paid, mister. Mycorrhizal fungi are far less likely to get on board when plants have excessive amounts of soluble ions to slurp up. Plants that would otherwise be responsive to AM fungi can indeed become fully resistant to colonization when bathed in MiracleGro or other commercial plant foods. Organic soil amendments, on the other hand, are primarily rock dusts and proteins sourced from plants and animals (such as soybean meal and crushed poultry feathers). These sources of slow-release nutrients should be thought of as food for microbes . . . which gets us heading back in the direction of a well-adjusted soil quite nicely.

Let's juxtapose the outreach of the common mycorrhizal network with what we learned earlier about the surface volume extension of the mycorrhizosphere. Hyphae effectively enlarge the absorbing capability of a single root system by ten to one hundred times. Let's ramp up those numbers to reflect networking reality. Researchers will readily say it's one thousand times more. Do I hear ten thousand? One hundred thousand? We don't actually know just how far a single mineral might get carried on an inbound journey, to be honest. Networking from root to root, with passage plants along the way, could encompass considerable territory. Roots also serve as a vehicle to carry fungal connection to ever greater depth.

Balanced nutrition is a two-way interaction where both plants and fungi sense and reward what the other does.

IMMUNE ALERTS

Plants certainly do communicate. The exchange of chemical signals in-house takes place by means of vascular transport of metabolites. Organic volatiles released by an affected leaf tell unaffected leaves of impending insect and pathogen attacks. These airborne signals prompt other plants in the immediate vicinity to mount defenses as well. The systemic compounds produced as a result of such messaging either repel the pests or attract beneficial organisms that can fight off the pests. The question is, can fungi spread a similar warning far and wide?

Mycorrhizal networks apparently do. Plant hormones and other signaling molecules are carried by means of fungal protoplasm to entire communities of plants. The jasmonic acid pathway (see "Induced Systemic Resistance," page 43) is the basis for this fungal-mediated interplant communication.[13] Think of hypha filaments as being like fiber-optic cables, carrying information between plants of the same or even of different species.

Induced defenses minimize the metabolism investment required of plants by eliciting a further response only should it prove necessary. Plants stand primed to engage the challenge once forewarned. Only this strategy takes time to engage, and thus there is value in being able to share signals across an ecosystem ahead of time. The common mycorrhizal network allows those plants located at some distance to "eavesdrop" on plant defense signals emanating from a plant experiencing some form of stress. From chomping caterpillars and drought to bacterial onslaught and rot. Researchers have found that signal molecules transferred through mycorrhizal networks elicit behavioral changes within twenty-four hours of an aphid infestation.[14] Speed is of the essence.

Electrical signals may play an important role in readying a proper state of plant preparedness even sooner. Wounded plants experience membrane depolarization at the cellular level. This shift from a negative to a positive internal cellular environment

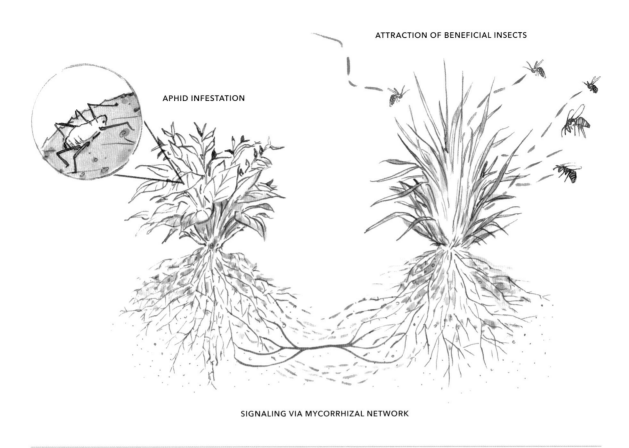

APHID INFESTATION

ATTRACTION OF BENEFICIAL INSECTS

SIGNALING VIA MYCORRHIZAL NETWORK

Messaging between plants by way of the mycorrhizal network consists of signal molecules and electrical pulsing.

allows for the transmission of electrical impulses both within a cell and, in certain instances, between cells. Can these "electrical action potentials" be detected and carried along by the mycorrhizal network as well? There are reasons to believe this just may be the case.

Various aspects of fungal physiology and activity have been shown to be affected by membrane depolarization, so involving it in interplant signaling isn't too much of a stretch.[15] An electromagnetic wave passing through fungal protoplasm seems entirely plausible. Sound waves travel more than four times faster through water than air. Similarly, fungal reverberations may in turn enhance the message by beating signal molecules to the punch.

Nondisturbance Principle

We approach reveling in a mycelial world in different ways. The one and only ground rule is that good *fungal infectivity* of the soil will always be contingent on minimal disturbance of the soil. Whatever you are doing now you can undoubtedly do better. Catch this fire in your belly and you will be ready to work with energy and determination to keep things humming for a long time to come.

Biological Compromise

Not every situation lends itself to going fungal from day one. A human comes along with healing intentions. Degraded soils shout out *Whoa, Jack.* Genuine structural needs must often be addressed first in the process of getting to lasting balance.

And this is totally fine. Soil prep for an orchard entails practical ecosystem conversion along with bringing cation balance down to the subsoil layer. Swales formed across rolling corn ground capture water for an envisioned agroforestry planting . . . but only if heavy equipment moves some earth to make this possible in the first place. Building good tilth from compacted ground requires a changing of the guard so that cover crops can work miracles. Briefly disobeying the inherent tenets of Organic Matter 101 by using black plastic to knock back rhizome-spreading grasses does not send you to purgatory (*though this will certainly call for ten Hail Marys, my son*). Going slightly allopathic in transitional situations to deter disease until healthy plant metabolism comes up to snuff might indeed be what pays the mortgage in those critical first years.

This is admittedly hard. Rigid adherence to an assumed ideology always seems so straightforward. Allow yourself this degree of biological compromise. The fungi can wait just a wee bit more now that you are at the helm.

Begin with gratefulness. Everything we have, everything that we know and love, is a gift. The soil we crumble between our fingers; the fungi that made it so; the carbon that marks our every breath. Profound scientific discoveries and the practical applications that follow flow best from a gracious heart. We have been given enough for every need and this can be shared. Generously. Like the fungi do for plants. Like the plants do for fungi.

Don't screw up. Talk about getting right back to the issue at hand! Well-intentioned people have done many things that aren't necessarily good for the fungal realm, myself included. More often than not we have treated the land like dirt — that's what the soil would be without soil life. Atoning now and then isn't such a bad thing. What counts is continuously recognizing how Nature goes about health and then adapting those hard-won lessons. Some of us won't be receptive to fungal guidance no matter how it's presented. No offense intended, but certain things need to be said:

Fungicides, herbicides, and insecticides no longer have a place in agriculture or forestry. Chemicals proved alluring and mistakes were made. Moving forward with a fungal mindset shifts the perspective to appropriately judge such

64

These zucchini plants in a no-till garden plot abound with fungal health.

matters. Biocides harm beneficial organisms in the root zone and on the foliage. Herbicides and fungicides hit mycorrhizal fungi especially hard. Healthy plant metabolism points to far better ways to grow nutrient-dense foods.

Synthetic nitrogen fertilizers are directly tied to the decline of soil carbon. One ton of applied nitrogen results in nitrogen-feeding bacteria consuming 30 tons of carbon. Oxidation of freshly deposited carbon compounds (when these bacteria are mineralized) means most of this carbon will be released back to the atmosphere. We need to favor species in the soil food web that build humus. Like fungi. And more fungi. Score another one for biological farming.

Tillage reduces the efficacy of mycorrhizae by disrupting the extraradical hyphal network. All sorts of trouble follow from there. We will be discussing strategies to break ground open far less often, both in the garden and across acreage. Occasional tillage done wisely has a place — maybe once every few years — but doing so every year sets back the mycorrhizal system again and again. Learn to work soil with well-timed cover crops and biological decomposition instead.

Conventional monoculture needs to be seriously reconsidered. Fungal diversity coexists with plant diversity. We can have *amber waves of grain* but do so in the context of multiyear rotations, cover crop cocktails, and up-and-coming perennial strains, please.

Animal husbandry needs to be an integral part of a long-range fertility strategy on diversified farms once again. Vast amounts of soil carbon are safeguarded when grazing is done right.

Farm consolidation to the detriment of vital rural communities was a big mistake. Many more farmers working reasonable acreage would be a very good thing indeed.

The mad destruction of forests worldwide needs to stop. Look not just to the trees but to the soil below. A forest ecosystem has anywhere from less than ten species of ectomycorrhizal fungi to more than two hundred species. Driving countless species to extinction for toilet paper is not okay. Development of distinctive plant communities follows along altitudinal or latitudinal gradients . . . and we need every last one to thrive. Sustainable forestry must be based on a big-picture view of ecosystems and forest cycles that reach far beyond our own lifetimes.

The slug on this chanterelle mushroom will be distributing spores on its onward journey. Small matters hold greater portent than we often suspect. Photo by David Spahr.

Do fungal things. Good stewardship requires we be fully aware of fungal dynamics. There are so many caring ways to garden and farm and work in the orchard and woods. Time to accentuate the positive! From creating outrageous diversity and perennial-based systems to growing lush cover crops and spreading meaningful mulch. Be all about fungal health and greater aggregate stability. Reduce erosion while enhancing nutrient availability. Inoculate where needed. Cultivate gently. Plant more trees. Encourage vegetative cover year-round. Walk softly in the forest. Spread compost to the horizon. Incorporate biochar. Make every day a mycelial dance. And always think fungally whatever you do.

Honor the earth. We are a part of this natural world. Working to increase soil health is downright exhilarating. Getting one's hands in soil teeming with life is the path to discovering ourselves. *We are stardust, we are golden, we are billion-year-old carbon, and we got to get ourselves back to the garden.*

So goes the nondisturbance principle, live and inspired. A functioning soil moves from bacterial confidence to fungal elegance over a period of time with different plant actors chiming in through the process. Fungi and bacteria differ in their responses to changes in management practices. We'll be walking through all sorts of "nondisturbance techniques" ahead to gain a fuller sense of how best to promote mycorrhizal ascendancy in soil ecosystems. Fungi are far more sensitive to change than bacteria . . . which is what makes the fungal-to-bacterial-biomass ratio a good indicator of environmental impacts in the soil. This is what people are talking about when they refer to *fungal dominance* versus *bacterial dominance* in soils. Bioassays can be taken (much like testing for soil chemistry) but ultimately plant species composition and vigor tells the story just as succinctly.

Nature has all the time in the world to restore habitats by means of bacterial-to-fungal succession. Bare ground sees weedy species come, which in turn become fodder for first fungi. A modicum of mycelia allows perennial plants an improved niche, and thus the organic matter crescendo really begins. Woodsier plants grow, contributing cellulose and lignins. Successional tree species compete for space, with the losers going the way of fungal accrual. Biomass builds as diversity becomes a driving force in its own right. Old-growth forests come to be. Tallgrass prairies sway in the breeze. Savanna lands shelter grazing herds. All such "climax communities" become self-sustaining (more or less) through balanced mineral nutrition delivered by a fully functioning mycorrhizal network.[16]

The pace of all this doesn't really cut it for humans on a much shorter time line. Nor should it. Restoring degraded habitats needs tending to now. Growing food to feed ourselves requires soil investment in the form of copious amounts of organic matter. Soil ecosystems are made up of physical, mineral, biological, and energetic components. Conventional agriculture broaches the physical and mineral — you plow, you fertilize, you reap. Organic methods focus on natural inputs and begin to incorporate the biological realm with cover crops and compost. Biological farming puts yet greater emphasis on feeding microbes and surface decomposition. Permaculture seeks to bring everything round through diversity. Utilizing principles from all camps can have merit on the journey toward plant health and productivity. The only thing the fungi ask is that we make mycelial choices as best as we are able.

This particular riff on fall cover cropping sets up the spring garden nicely. Tillage radish, field pea, and oats all will winter kill, leaving reliable mulch behind in which to transplant. The radishes drill deep, aerating the soil without any mechanical disturbance. Peas and oats carry mycorrhizal fungal connections forward. This sort of *biological tillage* carries out the nondisturbance principle to perfection!

Let's end this with some advice from a dear friend who has crossed the divide and yet speaks as kindly and sagely through his poetry as if it were yesterday:

> *Mostly I think about how important it is to hold*
> *and care for what is still here. Two hundred years ago*
> *it was easy to plant and feel a sense of abundance,*
> *replacing giant pines and maple with annual grains*
> *and hay crops. But that fertility was used hard,*
> *and not replaced, and that means the job for me,*
> *for you, for those to follow — will be to capture*
> *as much as we are able,*
> *to cycle the leaves into soil, into food,*
> *to hold onto what lies beneath.*

> RUSSELL LIBBY, AN EXCERPT FROM
> WHAT YOU SHOULD KNOW

CHAPTER 4

Provisioning the Mycorrhizosphere

The root zone has been described as being an intimate association of plant and soil life, to the point that it can be difficult to separate *what is plant* from *what is microbe*. Concepts of continuity work for the best in thinking about rhizosphere digestion and soil stability. We're expanding now on that idea of the fungus-root as a team by evoking all the players involved. Both fungi and bacteria colonize root tissues in the mycorrhizosphere. The bulk soil beyond—yet to be affected by the active functioning of roots—will ultimately be incorporated as well, as roots and fungi alike explore new spaces.

It would be a very different story without these biological thrusts. Soils experience a remarkable set of transformations over time, as energy, chemical elements, and water get continually sorted. Primary minerals become weathered and wash away. Soil acidity increases due to atmospheric deposition. The loss of exchangeable nutrients through agriculture and whole-tree harvesting only adds to the void. Soil weathering over eons can potentially exhaust all primary minerals and a number of chemical elements to ever greater depth, leaving behind only the most insoluble chemical elements and recalcitrant minerals.

The spiral of entropy does not go unchecked, obviously, as life renews the soil by means of organic matter from the top down.

Time for action! Mycorrhizae play a preeminent role in structurally stabilizing soils. This in turn provides a home base for further nutrient reach to support healthy plant metabolism. We sustain this all the more by providing "fungal foods" that the right sorts of soil organisms thrive on. The pivotal question of mineral investment considers if the soil food web alone can suffice in every situation. We'll also look at "water rights" as understood by plants and fungi alike. Providing the occasional nudge toward abundance works for me.

The Glomalin Connection

Arbuscular mycorrhizal fungi and feeder roots reach out into untapped nutrient zones throughout much of the year. Each foray occurs in a span of mere days to as much as a few weeks. A portion of hyphae and finer roots are shed as plant and fungi alike invest in new directions. Residual organic matter is quickly consumed by numerous microorganisms as the great wheel of assimilation and mineralization turns again and again. Yet this continuous ebb and flow of hyphae leaves an enduring remnant of its passing every time.

Fungal hyphae are fundamentally open conduits transporting water and dissolved minerals back to host plants. The cellular coating of these hairlike filaments needs to be relatively leak-proof. Accordingly, arbuscular mycorrhizae secrete a gooey glycoprotein known as *glomalin* to seal off intercellular spaces and give hyphae some rigidity to push onward through air spaces between soil particles. This carbon-rich glomalin gets left behind in the surrounding soil when the mycelium attunes to new horizons. There it will stay for anywhere from seven to forty-two years, depending on conditions, as the superglue that holds soil aggregates together.

Glomalin eluded detection until the 1990s because this molecule of glycoproteins holds fast so effectively. USDA soil microbiologist Sara Wright named glomalin after the Glomerales group of arbuscular fungi. What has stunned soil ecologists ever since is that glomalin contains as much as 40 percent carbon in its molecular structure — representing nearly a third of carbon storage in soils worldwide. Humic substances, by comparison, hold only about 8 percent of soil carbon reserves.

A resounding link between soil biology and carbon sequestration now stands revealed.[1]

Furthermore, higher carbon dioxide (CO_2) levels in the atmosphere will stimulate fungi to produce more glomalin. Multiyear studies done in California revealed that when atmospheric CO_2 reached 670 parts per million (predicted to be coming our way by 2050), mycorrhizal hyphae grew three times as long and produced five times as much glomalin as fungi on plants growing with today's ambient level of 370 ppm.[2] Longer hyphae help plants reach more water and nutrients, which will help plants face drought in a warmer climate. Managing soils biologically makes far greater sense than pushing our existence to the extreme, of course, but at least fungi have a backup plan.

Glomalin's role in holding soil particles together kicks off the provisioning of the mycorrhizosphere. Good soil structure can be gauged by the preponderance of soil aggregates. Loose sand, silt, and clay particles glomp together through wetting and drying, freezing and thawing, and by root growth and earthworm activity. The initial holding power between negatively charged soil particles and positively charged ions of calcium and magnesium is best described as an association that then embeds bits of organic matter as well. Glomalin "cements the deal" and thereby gives aggregates the ability to hold together in both wet and dry conditions.[3] Earthworm-created aggregates (worm castings) are water-stable from out of the worm. More of everything gets pulled into the developing structure, exactly like a snowball picking up mass as it rolls down a slope. Smaller aggregates become bound with one another to create macroaggregates.

Nutrient exchange doesn't get any better than in these mycorrhizal havens, where a porous structure invites microscopic hyphae to explore freely. This increases glomalin potential within, making

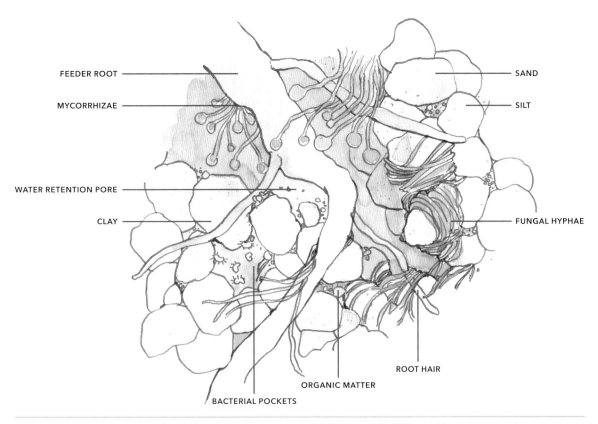

FEEDER ROOT

MYCORRHIZAE

WATER RETENTION PORE

CLAY

SAND

SILT

FUNGAL HYPHAE

ROOT HAIR

ORGANIC MATTER

BACTERIAL POCKETS

Soil aggregates are the result of good fungal stewardship. Hyphae will abide within, delivering fertility in long-lasting fashion.

aggregates ever stronger. Glomalin is not exceptionally recalcitrant in its own right, but becomes more so as part of a package deal. Openings into a soil aggregate are too small for most bacterial and fungal predators, thereby creating a "gated community" for mycorrhizal fungi and helper bacteria. This leaves the nutrient scene almost exclusively to the wiles of fungal hyphae. The house that mycorrhizae have built — the soil macroaggregate — benefits mycorrhizal fungi most of all. Feeder roots are certainly too big to penetrate directly into this scene and so must rely on carbon negotiations with the developers.

A well-aggregated soil has increased aeration and water-holding capacity. Plant roots stretch out

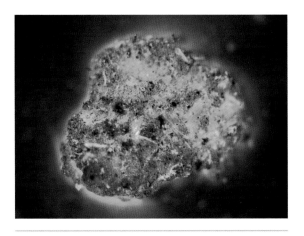

Carbon-rich glomalin helps bind this soil aggregate into a fertile whole. The fluorescent-green dye used to make this image suggests where the true emeralds of this world lie. Photo by Kristine Nichols, Rodale Institute.

all the more in a soil that's high in organic matter, thus occupying a greater share of soil volume. Earthworms and macroarthropods move about easily while distributing nutrients anew. Surface crusting simply doesn't occur when biology has the reins in hand. The soil as a whole proves more erosion-resistant. Soil carbon that has been "stabilized" (see "Carbon Pathways," page 99) includes the glomalin fraction. Arbuscular mycorrhizae fungi have established a home base from which to seek the next tasty morsel.

Fungal Foods

That saprotrophic fungi and mycorrhizal fungi find sustenance in organic matter at different stages and for seemingly different purposes is more a division of function than a separation. Fungi need nutrients as much as plants do. What might be called *fungal foods* in the end are as much *plant foods*. This way of speaking need not confuse anyone.

Nor do growers truly "feed" fungi. A mycorrhizal spore germinates and its fledgling hypha attaches to a root — the journey is a fait accompli. That leaf fallen to the ground will disappear despite us. Plant productivity, on the other hand, benefits from a rich and diverse nutrient base. This becomes all the more fulfilling by recognizing that certain types of organic matter jump-start life. Understanding a bit of the workings behind soluble lignins, fatty acids, and bacterial metabolites empowers our work.

SOLUBLE LIGNINS

A branch falls from a tree; crowded saplings succumb to succession. Lignins now lie on the ground in whole branch form.[4] Time passes as assorted fungi go to work breaking down this woody organic matter.

Lignins mainly occur in the intercellular layers of the cambium of woody and nonwoody higher plants with a vascular system. From grasses on up to the towering redwoods, lignins provide strength and rigidity to xylem cell walls found in the cambium tissues of roots, stems, blades, leaf veins, shoots, limbs, and woody trunks. All that lignin biomass constitutes 30 percent of the organic carbon found across terrestrial ecosystems — thereby imparting lignin with great significance as a source of lasting soil fertility.

That job calls for particular wood-rotting fungi able to do right by that carbon.[5]

One group of such fungi concentrates exclusively on softwoods. The *brown rots* consume cellulose, leaving behind a darker lignin, which eventually dries and cracks in a cubical fashion. Brown rot species are relatively few and lack enzymatic mechanisms to decompose lignin completely. Compounds produced by these fungi in the early stages of decay act to suppress deciduous species where conifers have established a niche.[6] Brown rot residues end up being highly enriched in recalcitrant lignin, resistant to further decomposition.

Another group of fungi tackle both lignin and cellulose. Several thousand species of these *white rots* have been identified. Some white rots preferentially remove lignin from wood, leaving behind bleached pockets of degraded cells consisting entirely of cellulose. Others break down lignin and cellulose simultaneously. The rate of wood decay depends on the assorted extracellular enzymes produced expressly by these fungi.[7] White rot fungi are found predominantly on hardwoods, but will colonize softwoods on occasion. These saprotrophs create long-term fertility in the form of fulvic and humic acids, assumed by some to be the building blocks of humus.

The formation of humic substances is one of the least understood aspects of lignin decomposition. Things are turning right controversial (see the humic mysteries section within "Carbon Pathways," page 101), yet all begins with a fungal-driven process. Forest-derived soils come at this through a speculative pathway that begins with white rots making carbon available. One explanation holds that polyphenol compounds released from lignin during this microbiological onslaught undergo further enzymatic conversion to then be combined with nitrogen-rich amino acids, resulting in what might best be called humic polymers. This "dark earth" is what all gardeners, orchardists, and farmers strive to develop through holistic management of soils.

The lignin of hardwood trees is central to building humus in forest-derived soils. Angiosperms (hardwoods) contain both guaiacyl and syringyl lignin in varying ratios among species, while gymnosperms (softwoods) contain primarily guaiacyl lignin. The enzymatic chemistry of the white rots works best with deciduous wood, whereas the non-enzymatic chemistry of the brown rots is another carbon journey entirely.[8]

Lignin becomes all the more transmutable by white rot fungi when soluble. The newest growth of trees — represented by buds, shoots, and twigs — contains a less polymerized form of lignin. That the focal point of growth energy in trees can be found in the buds and supple bark tissues of twigs exposed to full sunlight makes sense. Nutrient levels run high where cambium cells directly supply the photosynthesis process. All told, 75 percent of the minerals, amino acids, proteins, phytohormones, and enzymes found in the tree turn out to be in this smaller wood.

The soluble lignins found in freshly chipped wood from the twiggy branches of deciduous species rock the biological kasbah.

We have come to the land of opportunity for building rich soil.

Back in the late 1970s and 1980s, researchers at Laval University in Quebec asked a pivotal question: *What can be done with the tops of trees resulting from logging operations for agricultural purpose?*[9] Obviously smaller wood could be left in the forest to renew those soils. Yet think about the untold amounts of lignin-rich organic matter resulting from clearing pasture and along power lines, pruning suburban landscapes and fruit orchards, selective thinning in woodlots, even deliberate coppicing of noncommercial species such as alder and poplar. All locally resourced and essentially free, with the soluble lignin advantage to boot!

Ramial Wood Nuance

The attributes of the right mulch "ramially speaking" spring straight from our lignin discussion:

Species for chipping should mostly be deciduous, as here lies the soluble lignin advantage.

Up to 20 percent of the mix can be softwood species without altering fungal balance.

Dormant wood (without leaves) is preferable. Leaves in full green bring a higher nitrogen content and thus greater bacterial influence.

RCW consists of twig wood less than 7 centimeters in diameter — not much more than 2½ inches around at the large end of the branch.

The proportion of essential twig nutrients increases as average branch diameter decreases. Nitrogen, phosphorus, potassium, calcium, and magnesium are found in the green cambium cells where leaf photosynthesis and root nutrition come together to make a tree . . . which we in turn can redirect to build ideal soil.

The carbon-to-nitrogen ratios in ramial-diameter wood averages 30:1, going no higher than 170:1 at the larger end of the recommended branch. These ratios rise dramatically in stem wood, running 400:1 to as much as 750:1.

Coarse pieces are preferable to finely shredded mulch (which overaccelerates soil digestion of RCW).

Suitable woody material, once chipped into smaller pieces, becomes *bois raméal fragmenté* in French. Here in the States, *ramial chipped wood* (RCW) is taking on epic stature as the go-to mulch for building soil. Another term you might hear is *arborist wood chips*, referring to the trimmings sought after from landscaping companies.

Finding sources of "soluble lignin" has gotten quite competitive as more people tune into the fungal food aspects of ramial wood chips. First and foremost is that smaller wood from deciduous trees eventually gets transformed into soil aggregates and highly reactive humus by the soil food web. Mycorrhizal fungi in turn utilize lignin-derived

macroaggregates as a center of operations for directing more complex forms of nutrition onward to plant partners. Gardens and farms, forests and orchards merit more of this! The ins and out of applying ramial wood nuance lie ahead (throughout chapter 6) when we explore how to shift soils toward greater fungal embrace.

Fatty Acids

One grouping of fungal foods should be considered essential, as it's through fatty acid nutrition that growers can quickly further plant and microbe health.

Yet first let's think about you and me. Go to the health food store, should you be so inclined, and chances are one of the things on your Be Outrageously Healthy List will be a fish oil supplement. The point here is that omega-rich fatty acids are necessary for human metabolism.[10] The kind our ancestors speared. We may not be thinking about linoleic and alpha-linolenic acids per se, but it's these unsaturated fatty acids and others on which cellular biology thrives.

All life-forms make use of lipids, proteins, carbohydrates, and minerals to fuel the system. Let's steer this in a plant-fungal direction and spotlight lipids. Fatty food components rarely get talked about in horticulture or mycology circles. And yet here lie some scintillating revelations for holistic growers with a fungal bent.[11]

Lipids may be relatively simple molecules, such as the fatty acids themselves, or more complex and contain phospho or sulpho groups, amino acids, peptides, sugars, and even oligiosaccaharides. Plants produce the majority of the world's lipids, and all animals, including humans, depend on these fat sources of calories as a vital part of a healthy diet. Like other organisms with visible nuclei, plants require lipids for cell membrane synthesis, as signal molecules, and to store carbon and energy. Additionally, leaf tissues and bark have distinctive protective lipids that help prevent desiccation and infection.[12]

Fatty acids serve as core components within all these assorted lipid molecules. Fats, waxes, sterols, fat-soluble vitamins (such as vitamins A, D, E, and K), monoglycerides, diglycerides, triglycerides, and phospholipids share an unbranched carbon structure and are insoluble in water.[13] It's the bonding mechanisms between carbon atoms that bring us back to more familiar language. If there are no double bonds between carbon atoms, the fatty acid

Table 4.1. Fatty Acid Classifications

Fatty Acid	Classification
Linoleic acid	Polyunsaturated
Oleic acid	Monounsaturated
Palmitic acid	Saturated
Stearic acid	Saturated
Alpha-linolenic acid	Polyunsaturated
and many more!	

is *saturated*; if there are double bonds between carbon atoms, the fatty acid is either *monounsaturated* or *polyunsaturated*.

Fungi produce a variety of lipids featuring distinct profiles. Linoleic, oleic, and palmitic acids are the main fatty acids in most cases. Linoleic acid is very abundant in fruiting bodies and spores, accounting for as much as 70–80 percent of the lipids present. Levels of oleic acid rise in mycelium to more equal proportion. These unsaturated fatty acids help basidiomycetes adapt to low growth temperatures and may have a role in lignin degradation. Arbuscular mycorrhizal fungi rely on transfer of palmitic acid from the root cells of its host.[14] Saturated fatty acids taken up by intraradical mycelium become the means by which AM fungi store hard-earned carbon and energy from plant sugars.

Another revealing twist on the fatty acid highway comes by way of the work of microbial ecologists. Scientists use fatty acids as a chemotaxonomic marker of specific fungal and bacterial presence in soils. Phospholipids are the primary lipids composing cellular membranes and thus used as the indicator molecule. Phospholipid-derived fatty acids are released when the lipids in an organism sample are saponified.[15] The composition of the resulting fatty acids can be compared

Goodbye Fungicides

Chemicals designed to kill pathogenic fungi in order to protect crops from disease are called *fungicides*. A reasonable person might ask if use of such would be bad for other fungi. Both the soil and arboreal food webs are rich in species that help plants be healthy . . . all of which take a hit when human beings indiscriminately medicate by focusing attention on the needs of unhealthy plants instead.

Industry research shows that mycorrhizal fungi can be quite sensitive to some fungicides, but not necessarily all. The preponderance of evidence indicates that the following classes are likely detrimental to mycorrhizal colonization and survival: aromatic hydrocarbons, copper compounds, imidazoles, phthalimides, strobilurins, thiadiazoles, and thiazoles. Foliar applications of nonsystemic fungicides are considered relatively harmless. Combined use of multiple types of fungicide, however, will "amplify negative effects" (and I indeed quote). Soil fumigation brings death to symbiotic anything for a long, long, time.

The far better approach is to work with healthy plants. Fungal foods, diverse polycultures, competitive colonization on the leaf surface, investment in minerals — most certainly accompanied by bountiful mycorrhizal stimulation of plant defenses — brings about the sorts of crop productivity that really matter.

The work of Alyson Mitchell and others at University of California–Davis on antioxidant levels in fruits and vegetables tied to growing methods reveals that fungicide use reduces important phytochemistry in the foods we eat by a factor of three. That venerable saying about an apple a day keeping the doctor away shifts to many apples a day when our growing methods are not health-based but rather exclusively focused on dealing with symptoms.

Pass the fatty acids, please.

with recognized lipid profiles of organisms already in the database to determine the identity of the sample organism.[16]

All of which suggests that fungi and plants are not especially keen on a low-fat diet.[17] Naturally enough, it's time to go full-spectrum in considering foodstuffs that benefit soil fungi as well as arboreal dynamics on foliage and bark surfaces. Sources rich in linoleic, oleic, palmitic, and stearic acids (and to a lesser degree alpha-linolenic, palmitoleic, and other acids) make the list for fatty acid application in gardens and orchards.

My use of fatty acids as a plant-fungal catalyst commenced with grapefruit seeds. It turns out that citrus pips in general are loaded with useful fatty acids. The whole show, in fact. Back in the 1990s, a grapefruit seed extract prized for horticultural purposes was being marketed for tropical crops.[18] Thanks to a tip from a fellow apple grower in Michigan, I launched into the fatty acid scene.

Doug Murray also turned me on to another seed oil from India that came with the full promise of a triple whammy.

Pure neem oil. The nutrient-rich oil pressed from the seeds of the *Azadirachta indica* tree offers three areas of intrigue for holistic growers. Terpenoids in neem stimulate an immune response to ward off fungal disease. Azadirachtin compounds inhibit the molting cycle of pest insects. Finally, a generous array of fatty acids in neem serves as a fungal feed for the soil organisms as well as arboreal organisms.[19]

Neem has a long history of use in Ayurvedic medicine and agriculture. Those centuries of tradition come based on *whole plant medicine* principles that herbalists the world over espouse. Therefore it's important to find a source of unadulterated, cold-processed, 100 percent, never frigged, pure neem oil.[20] Other neem wannabes on the market are extracts of the whole. The constituents removed first are the fatty acids, for the transgression of being inconvenient to spray. Way to think things through, boys and girls! Avoid those patented neem products like the plague. The raw seed oil tastes bitter and smells garlicky, if not nutty at its freshest.

Fish hydrolysate. Liquid fish, technically known as *fish hydrolysate*, is made from the first pressing of genuine fish parts and has not been pasteurized — and thus contains the expected fatty acids and enzymes. The lipid profile of fish includes omega-3s.[21] Recommended brands in North America include Organic Gem, Neptune's Harvest, Native Nutrients, Dramm, and Schafer Fisheries. These companies use an enzymatic, low-heat process to ensure that organic compounds are left intact

while eliminating bacterial breakdown (and thus strong odor) by adding a trace amount of citrus extract or phosphoric acid. These methods result in fatty acids ready for dispersal.

Fish emulsion, however, consists of liquid wastes (after processing fish for other purposes) that have been heat-treated and thus biologically deactivated. Fish remnants are cooked in order to remove the oil portion for paints and cosmetics. Then the protein is removed and dried to make fish meal for livestock feeds. The remaining wastewater is condensed by half into a brown, thick liquid called an *emulsion*, which has considerably less nutritive value than true *liquid fish*.[22]

Understanding qualitative aspects makes all the difference when it comes to fatty acids.

Whole milk. Grass-fed cow's milk comes from cows that have grazed in pasture rather than being brought a processed diet of grain and silage inside the barn. Green grass makes the milk richer in omega-3s, vitamin E, beta-carotene, and a beneficial fatty acid named *conjugated linoleic acid*.[23] It goes without saying that whole milk delivers more fatty acids than skim milk. Milk from a cow contains roughly two-thirds saturated fats, a tad of polyunsaturated fats, and a remaining "short third" of monounsaturated fats. Pasteurization has no effect on fatty acid composition in milk.[24] Raw milk is superior in vitamin and mineral content overall . . . but, as far as fatty acids go, heat treatment is not a game changer.

Milk provides a foliar calcium boost to plants. Calcium has been shown to inhibit disease spore germination and thus effectiveness against powdery mildew fungi. Taking full advantage of the fats in whole milk (in addition to the calcium) requires lactic acid fermentation in an anaerobic brew barrel. This differs considerably from culturing probiotic yogurt, and yet depends just as much on microbes to prepare milk fats for improved absorption.

––––––––––

Vegan alternatives. Liquid fish by any stretch of the imagination is not a plant-based remedy. Nor for that matter are dairy products . . . though one could argue that just the day prior green grass made its digestive way to becoming milk through the rumen of the cow. Folks who want vegan alternatives have a number of options to try in holistic sprays, still based on fatty acid composition, with the same questions of cost and availability.

Coconut milk or further refined coconut oil first comes to mind, though given the buzz for this in regard to human health, maybe it's not the most economical choice. Coconut brings lauric acid into the picture . . . which may have little relevance to plant health whatsoever. Please accept this humble admission: *There be things we do not know.* Palm kernel oil used in making soap can be had for relatively little money, but this comes at great expense to the rain forests in Southeast Asia being cleared to make way for palm plantations. Hemp seed oil

Seed of the hemp plant has great potential as a fatty acid source for holistic agriculture. Photo by Alyssa Erickson, Kentucky Hempsters.

is exceptionally rich in essential omega fatty acids and proteins. Karanja oil, pressed from the nuts of Asiatic beech trees, has become a benchmark addition in my orchard sprays. This tree seed oil brings flavonoids into an exponential pairing with the terpenoids in pure neem oil to boost green immune function.[25]

Delivering these "good fats" to plant surfaces and open ground in order to spur fungal activity will take us to the tip of a spray nozzle. No worries on that score: The insoluble can indeed be made soluble. The molecular structure of a fatty acid includes a fatty chain on one end and an acid group on the other. The fatty end is *hydrophobic* (water-hating) while the acid end is *hydrophilic* (water-loving).[26] Breaking up this scene can be done through emulsification or some combination of hydrolytic enzymes and fermentation. Neem oil mixed with a small amount of a biodegradable liquid soap will subsequently mix into water (see "Emulsifying Neem Oil," page 125) because the neem has been emulsified. The fats in milk become fully bioavailable through lactic acid fermentation (see page 84 just ahead) in a plant-based liquid extract. Knowing tricks like this sets up success.

And while we certainly could include rendered pork lard and the like in this discussion of fatty fungal foods, these efforts need to be doable. Hard fats from animal sources are too greasy for spray application. Still, using resources on hand does have appeal. I was once asked in a class on holistic disease management about a more land-locked option to liquid fish. I have no idea if "liquid chicken" ever proved effective for the woman who asked[27] . . . but I sure do applaud homegrown ingenuity in applying constituent thinking to finding ever more local solutions.

BACTERIAL METABOLITES

Fungi get most jazzed when bacteria join the band. Among the ultimate fungal foods are nutrients released through extracellular digestion. Microbial jam sessions result in a riff of stimulatory volatiles and phytohormones as well from bacterial involvement. In one sense, such "bacterial metabolites" are intended for bacterial purposes — normal growth structure and energy storage, stimulating breakdown enzymes and catalytic activity, even defense signaling — yet fungi can and do partake for fungal benefit. *Hey, it's a party!* The more diverse the food web, the more diverse the fungal diet.

Growers can contribute to this crescendo in one of three ways.

The overarching creed of organic gardeners is compost and more compost. Yet far more is afoot than merely returning nutrients from organic matter to soils. Compost brings a diversity of species onto the playing field. Indigenous microbes live right next to plant roots . . . so add a few shovelfuls of topsoil now and then to build a mixed pile.[28] Earthworms introduce microbes from a very meandering gut biota. Animal manure of any sort ups the ante of extracellular enzymes to be shared. Mesophilic and thermophilic bacteria with assigned missions join this ebb and flow as the pile heats and then cools in the process of active decomposition. Plus the actinomycetes. Plus the saprotrophic fungi. Plus the flagellates and ciliates. Plus the nematodes. All contributing nutrients to do it all again.

Spreading compost should really be seen as spreading life across the land.

Ground sprays of compost tea are a brilliant means to further distribute diversity. The life composition of these brews begins with the beneficial organisms in compost (or vermicompost), which multiply by the millions and millions in a deliberate aeration process. Air bubbles and slight warming throughout the activation phase keep the tempo lively. Foods like fish hydrolysate, blackstrap molasses, kelp meal, and humates added to the tea get things going. Protozoa feed on the resulting bacterial flush, thereby revving the tea into high gear, with microbes multiplying by the billions and billions. Fungal hyphae extracted from the compost keep microbe representation on a complementary footing. A cup of powdered oatmeal per gallon of compost goes far in feeding the fungal element of a compost tea.

Ground sprays that include effective microbes introduce a facultative array of organisms. Holistic applications qualify here. The lactic acid bacteria in these probiotic cultures help further the decomposition of organic matter on the soil surface. Activated effective microbes combined with fatty acids in turn get soils churning out ever more bacterial metabolites upon which all fungi thrive.

Infusing the soil with organisms effectively arouses the *chi* (life force) of biological cycles.

Lastly, and fine-tuned at that, are particular organism products chosen to inoculate stressed and depleted soils. Bacterial inoculants can be used to make sure requisite organisms are on hand in those situations where human activity has undermined soil health connections. And there are plenty of those! Newly planted trees especially benefit from a bacterial boost until the fungal system is up and running. Bacterial metabolites turn out to be a foundational fungal food in these scenarios.

A number of regional suppliers have proprietary formulations of minerals, proteins, and humates that include beneficial bacteria. All fine products with promises aimed at increasing root biomass, decomposing field stubble, making essential nutrients bioavailable, and increasing levels of plant growth hormones. More than likely, the good folks at Tainio Biologicals in Washington State have something to do with culturing the microbe aspect of these foliar and field spray formulations. Many biological products contain only one or two strains of bacteria, targeting very specific criteria. Tainio creates microbe blends that are far more wide-ranging in synergy and function.[29] Like Nature herself, placing the emphasis on diversity.

Let's do roll call.

> *Arthrobacter globiformis.* Here.
> *Azospirillum lipoferum.* Yo, bro.
> *Azotobacter chroococcum, Azotobacter vinelandii.* Both here.
> *Bacillus amyloliquefaciens.* Yo' mamma.
> *Bacillus cereus.* Gotcha.
> *Bacillus megaterium.* Present.
> *Bacillus subtilis.* Where else?
> *Micrococcus luteus.* You and me, dude.
> *Pseudomonas fluorescens.* Here.

And on it goes.

Any mix of plant-growth-promoting bacteria can be combined with mycorrhizal inoculum into a single product. Bacteria play a key role in presymbiotic survival and growth of fledgling hyphae in the soil. Subsequently, nutrients are broken down through enzymatic action, facilitating easier uptake by the fungi and vice versa. The term *mycorrhizal helper bacteria* merely reflects this mutual dependence between symbiotic fungi and bacterial friends. Certain "myco blends" likely include *Trichoderma* species of fungi as well, which add an endophytic assist to the plant relationship.[30] Some people worry that these biocontrol fungi can set back mycorrhizal establishment, and thus do separate inoculations to allow symbiotic fungi a few weeks to develop before introducing the root endophytes.[31]

Mineral Investment

The attentiveness of pests and disease organisms to seek a stressed niche is a given. Maintaining healthy plant metabolism for whatever comes this way necessitates that ample mineral nutrition be available.

We kick this off by diving straight into the rabbit hole. *Whoosh.* Some argue that a functioning soil biota — restored, agile, and entirely diversified — can access all the minerals that plants need in any given soil, forever. Native plants in an undisturbed setting certainly speak to such naturally regenerative prospects. Score one for the biology camp. Others are convinced that good mineral nutrition requires a human touch. That a soil derived from granite bedrock, for instance, does not necessarily have every essential element in proper proportion. That those centuries of cropping since the last glacier passed by must be reckoned.[32] Saying all soils should be good to go flies in the face of geological reason. Score one for the remineralization camp.

Biology propels soil health, but by no means will it always be sufficient in and of itself for ongoing crop production. Barren lands need to be revitalized first with frontier plants. Degraded soils require boosts of every sort. Organic cropping, even in the wisest of hands, may result in lowered levels of essential elements over time. Fungi and bacteria can access and move locked-up minerals around, certainly, but no critter makes missing elements from nothing. The right question for growers to ask is how to keep the cost of mineral investment reasonable, knowing that a healthy soil biota will eventually offer so much more.

Shall we begin with the highest bidder? My friend Eliza calls this "super-pricey input land" in referring to consultant plans where mineralization recommendations (along with specific biology boosts) show little regard for the pocketbook. Case in point comes from a cider orchard planted in extremely stressed soils where management sought advice from two different perspectives. The cost for soil primers, transplant solution, biweekly fertigation, and weekly foliar applications under the one plan came to just under nine hundred dollars per acre for the first year of treatment.[33] There was no way to spend that kind of money. Plan two called for a mycorrhizal root dip, basic cation balancing, a dusting of Azomite, several holistic sprays, and spreading available wood chips, at a cost closer to four hundred dollars an acre.

In the middle of the road, yeah. The first job is to get living systems back online. Start by establishing biological connection. Use the roots of cover crops to further that connection. Make compost. Involve yet more roots. Ferment homegrown brews. Search out local resources such as the soluble lignins in ramial chipped wood. Invest wisely in modest amounts of minerals when you are able.

I tend to be a generalist cook when it comes to this. Bulk needs revealed on a soil test are addressed in early days. A dusting of granite meal and Azomite as you build a pile-in-progress adds an array of minerals to compost. Granulated kelp sprinkled on alfalfa pellets for the sheep through the winter months comes back out as richer muck. Lengthy garlic beds get a bag of gypsum and Azomite each as rotations direct this mineral outlay garden-wide over the course of multiple years. Basalt rock dust doled out by the shovelful is a special gift to ongoing tree vitality. I don't test soils continuously, but rather check in every few years, more for that "pat on the back" than anything else.

Never let mathematical fascination with soil chemistry convince you that lab analysis somehow

Broad-leaved comfrey pulls minerals up from deeper below via its taproots while valerian creates a beneficial insect haven around this apple tree.

trumps soil biology. We need to be about creating conditions that foster biologically active soils, even when applying a credible recommendation.[34] Bringing soil nutrients into balance with slow-release amendments sets the stage for vibrant green growth. I look for specific *fertility ratios* between one element and another in evaluating soil test results, knowing these will reflect back over time in the form of healthy plant metabolism. The cation side of the equation involves the positively charged bases—calcium, magnesium, and potassium—which support high biological activity, nutrient exchange, and optimal water intake and soil aggregation. All this gets lost by those who consider pH to be the whole enchilada.[35] The correct form of lime matters when it comes to cation balance. The pH reading provided by a soil test represents a number in flux . . . especially once you comprehend the notion of *localized acidity* in a rhizosphere (see "Biological Reserves," page 37) where fungi play a significant role. Mycorrhizal fungi prefer slightly acidic soils in general. Fungal activity in alkaline soils begins to falter above a pH much higher than 8.[36] Whenever you start stressing too much about exacting pH levels in the bulk soil, keep in mind that both plants and mycorrhizal fungi can modify the pH of the rhizosphere to create a more tolerable environment.

Soil tests sometimes reveal a glaring deficiency. Keeping up with boron in the Northeast comes to mind, given the primary role for this micronutrient in plant defense systems. Similarly, I have sprayed dilute copper in my orchard specifically onto trunks and the soil surface in early spring to improve biosynthesis of lignins and make available an essential cofactor in the production of antimicrobial compounds.[37] Plants need sulfur to make complete proteins . . . levels of available sulfur get scrutiny here as well. Gypsum contains about 18 percent sulfur in the sulfate form, which is far less damaging to fungi than elemental sulfur.[38] Soils in other regions have "hot ticket minerals" of their own to keep on the watch list.

Still, very few nutrient deficiencies are absolute. Give the biology time to come up to speed. Many suspected deficiencies are functional, due to poor soil structure or lack of microbial diversity. Different mycorrhizal species offer different nutrient specialties. Let's follow that trail back to the surface: Greater plant diversity woven into an expanding mycorrhizal network reflects greater mineral availability. Are you seeing this? Nature consistently stacks the deck in life's favor. Spreading mineralized compost and woodsy organic matter now and then simply complements the ordinary work being done in a humming ecosystem every day. Like a leaf falling from way up high to the forest floor. Like a taprooted plant bringing minerals up from below. Mineral investment at its finest involves teamwork.

Which brings the circle round to the growing of food and medicine crops. Agriculture invites ecosystem adaptation as a matter of course.[39] We'd best be honest about this, as that whole tug-of-war between unfettered biology and minerals to the nth degree will now be brought to a stunning conclusion. This generalist cook is about to happily hitch up his spray rig.

What's known as "foliar feeding" gives growers a targeted approach to boost crop production. Only now we are going to enlighten the ionists among us and say unequivocally that this, too, is a biological act. Spraying the foliage surface with chelated minerals, fatty acids, and microbe reinforcement to up the nutritional ante in plants is about far more than what skeptics accept or not. Granted, this ion or that may enter into intercellular fluids by means of larger micropores on the undersides of leaves or transcuticular pores throughout the cuticle.[40] As long as minerals remain dissolved in solution . . . such can be pulled within by good ol' chemical diffusion. But is that really the sum of all things?

We recognize the authenticity of the soil food web. Nor do we doubt the ability of roots to absorb bacterial metabolites and more complex nutrients delivered by fungi. These same organisms create a competitive environment in the rhizosphere to protect roots from soilborne disease. Carrying that principle into the light of day provides holistic growers with a leg up on deterring pathogens in

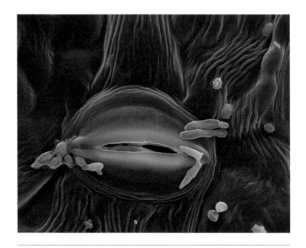

Nutrient uptake through stomata openings on the underside of leaves is more involved than most growers realize. Photo by Louisa Howard, Dartmouth Electron Microscope Facility.

Fermented Plant Extracts

Enlivened teas of nettle, horsetail, and comfrey provide homegrown sources of calcium and silica. It finally dawned on me to make "powerhouse brews" from these basic fermented plant extracts. Both the newly named *calcium tea* and *silica tea* come teeming with biological and nutritional wallop that radically improves overall plant health. Apple crops treated with these teas see limited sooty blotch and flyspeck, bitter pit, brooks spot, and other assorted spots come harvesttime due to a strong cuticle defense. Garden crops benefit from enhanced leaf vigor and increased productivity. These "recipes" can be tweaked to feature local plant resources . . . the main thing is to start brewing!

CALCIUM TEA INGREDIENTS

comfrey leaf
green nettle
effective microbes
whole milk (up to 5 gallons)
gypsum (calcium sulfate)
garlic scapes
humic powders

SILICA TEA INGREDIENTS

horsetail
seeded nettle
effective microbes
Azomite or soft rock phosphate
granite meal and/or basalt dust
humic powders

The basics of brewing plant extracts works as follows: Cut away the top of a 55-gallon plastic drum. Gather approximately 20 pounds of green herbs and loose-pack them into the drum. Fill the drum two-thirds of the way with unchlorinated water. Add 2 gallons of activated effective microbes, along with whole milk in the case of the calcium tea. Add 2 pounds each of available rock powders, stirring them into the solution at the top of the brew. Top off drum with water, using cut-off tops anchored by bricks to keep herbs in suspension. Stirring every other day or so helps enhance breakdown of rock powders and stimulate aerobic microbes. Fermentation lasts approximately ten to fourteen days; when complete, it is marked by a through breakdown of the herbs and an engrossing smell, to say the least. Roughly remove plant debris using a garden fork. Add ½ pound soluble humic/fulvic acids per drum. The brew will be ready for use, but can also stay in drum for subsequent sprays. A loose cover lessens evaporation.

Five gallons of each tea should be applied per acre in the orchard. Absolutely run these teas through strainers before adding to the spray tank. I use a coarse strainer for pouring tea into a 5-gallon bucket, which then gets poured through a fine-mesh strainer into the 100-gallon spray tank hitched to my tractor. Adding 3 cups of each tea in a backpack sprayer achieves the same 10 percent dilution for garden use. This may be a generous rate, but the goodness seems spot-on. A minimum dilution would be on the order of 3 percent, no less. Both brews are applied in the fruit-sizing window at seven- to ten-day intervals. I continue with the calcium tea in the fruit-ripening window at ten- to fourteen-day intervals up till harvest. These

get tank-mixed with seaweed, neem oil, and other ingredients in an ever-evolving holistic spray plan.

Garlic scapes aren't usually available for the first round of brewing the calcium tea, timed to be ready a week or so after blossom time in the orchard. (Scapes are the flowering stalk of hardneck garlic varieties, which are removed from the green plant when these

Silica tea with horsetail visible on top in the front barrel sits aside a batch of calcium tea featuring comfrey, green nettle, and whole milk.

The green vitality of a fresh stalk of nettle stands in stark contrast with the macerated herbs removed from the brew barrels. The goodness of those herbs has been transferred to the fermented plant extract.

Straining the teas removes plant particulates that would otherwise clog the pump filter on a sprayer.

begin to straighten in order to grow bigger garlic bulbs.) If you have remaining garlic bulbs in storage from the previous growing season, do use those up in this first round, for sure. The organo-sulfur compounds in garlic help facilitate constituent passage through a membrane. We are applying rich mineral brews to plant surfaces, aka *foliar feeding*, thus penetration into the waxy cuticle and indeed leaf/fruit cells is certainly an added bonus.

Some of you may want to add 5 to 10 pounds of compost or worm castings à la Jerry Brunetti to each batch. Epsom salts will benefit trees needing supplemental magnesium. Molasses feeds bacteria during the breakdown phase. The rules here are not hard and fast, other than adding humic and fulvic acid extracts in powdered form after the fermentation completes to help chelate released minerals.

Lastly, a sludge note. A goodly charge of the rock powders probably won't dissolve into solution. This keys to the fineness of the grind. Don't worry . . . any remaining mineral-rich sludge can be added to a compost pile, which eventually finds this investment in amendments returned to the soil.

the leaf canopy. Something is going on here . . . *it's at the tip of your tongue* . . . organisms above are capable of competitive colonization . . . *stick with this now* . . . that same organism community must also be able to assimilate and mineralize nutrients. Indeed. The time has come to fully embrace all aspects of the arboreal food web.

The diverse biota on plant surfaces includes numerous bacteria, single-celled yeasts, endophytic fungi stretching hyphae out from intercellular spaces within the plant, and even protozoa and the occasional nematode. All living and dying in three-quarter time. Nutrients taken in, nutrients released. The environment may have changed, but familiar dynamics still rule. Holistic sprays support this ongoing delivery system by providing nutrition and biological reinforcement in the canopy zone. So-called *foliar feeding* as a term doesn't come close to describing the full reality of what's taking place.

John Kempf speaks about critical points of influence (CPIs) in every plant's development. Active shoot growth, fruit-set, seed ripening, and meristem formation have more exacting nutritional requirements than times in between. Supplying trace minerals to serve as enzyme cofactors at these junctures enhances crop production in healthy ways. Foliar application reaches all parts of the plant more quickly than directing soluble ions into the root zone. Accordingly, I have started trialing MicroPak (Advancing Eco Agriculture's formulated blend of micronutrients) in my orchard sprays on either side of bloom. Boron, zinc, manganese, copper, cobalt, molybdenum, and sulfur are in a form that can be readily absorbed by both plants and the microbe community waiting on the surface. Time will reveal the gains to be had in terms of overall tree health.

Grower ability to time applications well and even make functional brews from indigenous

plants puts progressive mineral investment in a whole new light.

Moisture for Every Plant

One of the limiting factors for plant growth is water availability. Drought stress shuts down overall metabolism and reduces photosynthesis. This in turn impacts the flow of sugars to the roots and subsequent carbon trading. No self-respecting fungi are about to take this lying down.

Arbuscular mycorrhizae have a vested interest in reducing drought stress for plants. An expansive mycelium already brings water as a means of delivering dissolved minerals to the root system. Dry conditions trigger plants to close respiratory openings in leaves in order to reduce further moisture loss. Hyphal flow coupled with the "stomate valve" being stuck in the off position results in an increase in water pressure (turgor) in plant tissue. Wilting gets postponed, thereby supporting cell function for another day. Photosynthesis resumes and sugars are delivered. This explains why plants associated with arbuscular fungi outperform non-inoculated counterparts in regard to growth and mineral content during dry spells.

Moisture can also be shared between plants through mycorrhizal networks. Thus, water (and any nutrient resource being carried by that water) has the potential to move through multiple pathways to meet the greatest need. Such hydraulic redistribution benefits drought-intolerant plants established in an interconnected network of mycorrhizae with more drought-tolerant plants with deeper roots. Similarly, larger trees with far-reaching taproots can access water that's unavailable to establishing seedlings . . . and that water then gets redistributed at night through ectomycorrhizal mycelia to the young trees.

Let's follow this scene to the end of the deepest root. Ectomycorrhizal hyphae can extend as much as 12 feet (4 m) from the host root despite being very fine in diameter. This allows for geologic exploration into seemingly insubstantial crevices in *bedrock gone bad*. Porosity produced by subterranean weathering converts inert rock into a material that supplies organisms with habitat, stored water, and nutrients. Granite becomes what is known as *saprock* — a far more crumbly rock just as rich in minerals, with an extensive lattice of mesofractures that holds significant amounts of groundwater. Moisture tapped from capillary-sized pores soon begins the journey upward.

Biology makes possible restoration in remote areas where watering is out of the question. Eroded soils in the southern Sonoran Desert of Mexico are being brought back to life with a mycorrhizal assist. Researchers in Egypt have showed that inoculated pepper plants can survive and grow in the desert. Colombian farmers have used select strains of soil fungi to increase the yield of cassava plants by 20 percent on the savanna. And in India, mycorrhizae get full credit for enhancing drought tolerance in organic rice.

This is the right sort of green revolution, one focused on biology, with the promise of a far longer run than the failed chemical performance of the previous century. *Glomus deserticola. Glomus fasciculatum. Glomus mosseae.* These three arbuscular species are especially attuned when it comes to claiming water rights for plants. You'll often find at least two of the three in a quality inoculum blend. And don't let us forget the endophytic realm — an organic seed treatment called BioEnsure allows agricultural crops such as wheat and corn to withstand severe droughts and extreme temperatures. Adaptive Symbiotic Technologies works with an *endophytic fungus that shall not be named* (proprietary rights, ya know) discovered by means of watchful observation over twenty years, which can be applied to protect field crops naturally from abiotic stress.[41]

Perseverance and fungi. These are the gifts for our times.

CHAPTER 5

Fungal Accrual

Investing in species diversity begins underground. Questions abound when it comes to inoculating disturbed ground to restore biological health. Knowing more about who's who helps me appreciate the big picture of mycorrhizal networking. Soil-building work becomes all the more focused when we take into account how organic matter becomes stable carbon. Fungi lead the way in soil aggregation efforts regardless of how we might wish to interpret the humification debate that waits just ahead. Ultimately, what might be called *fungal accrual* is about creating the right conditions for fungi and thus green plants to thrive. The formula looks something like this:

*Organic matter in
equals fungal oomph back at ya!*

Inoculum Nuance

Some growers swear by inoculating with a mycorrhizal blend, while others expect "native spores" will always be sufficient to get the job done. Shades of gray aplenty. Let's spell out the knowns and speculate about the lesser knowns . . . and then you can decide for yourself what makes sense to do and what does not.

Mycorrhizal fungi occur in a wide range of soils throughout tropical and temperate regions. Which does not mean populations will be strong where there's history of chemical abuse, erosion, or other forms of soil disturbance. Cropping history ties in here, particularly after nonmycorrhizal plants have been grown. Natural dispersal of endomycorrhizal fungi into highly disturbed areas is slow, and the species that disperse more readily do not necessarily provide the essentials for every vascular plant. Those fungi that perform well in early stages of establishment may not be the best fungi for the long haul.

So goes complexity. Plants will always be receptive to fungal adaptability. The nutrient specialist that delivers zinc may not be the nutrient specialist that delivers manganese. Other fungi are adept at dealing with drought stress and saline soils. A vein of available nitrogen may well be tapped dry by the time those first silks emerge on corn . . . opening a niche for fungal assistance that wasn't there just days earlier. Mycorrhizal networking protocols will adjust to accommodate the needs of an evolving plant community. Climate change shifts plant and mycorrhizal topography alike. Diversity offers resiliency advantages as yet unseen.

Still, the right question to ask about mycorrhizal inoculation isn't about cost (and we'll get to that) but is about whether we are somehow screwing with natural systems inappropriately. Those who say a healthy soil already features indigenous species are right. Mycorrhizal populations prevailing at the present moment may well defend the home turf from introduced species . . . yet, chances are, the generalist species in a reputable commercial blend will share common ground with any existing fungal network and will contribute a diversity component at the very least. Arbuscular-type fungi are remarkably cosmopolitan, whereas this is not as often the case with ectomycorrhizal fungi.[1] You're about to get to know who's who. Let's focus on *less-than-healthy soil* as the operative phrase in deciding about fungal reserves.

Disturbance scenarios are foremost on the list for where biological remediation makes sense.[2] From strip mines to forests devastated by fire, from

The corn root ball on the left has far more root mass than its cohort on the right due to mycorrhizal inoculation. Courtesy of Graham Phillips, BioOrganics.

compacted fields to gardens long overtilled. The ins and outs of many such contexts lie ahead in the next chapter. Seedlings begun in a greenhouse in sterile potting mediums will always benefit from improved nutrient absorption and soil block aggregation. Medicinal herbs become more potent by means of mycorrhizal stimulation. Yields can be increased when fungi take hold on the majority of crop plants, saving on fertilizer costs. All trees and shrubs benefit from having friends on hand from that very first month in the ground. Lastly, there would be you. Merely contemplating this root investment tells me you're ready to foster fungal connections and *Grow Healthy*.

COMMERCIAL PRODUCTS

Mycorrhizal inoculants come in powder, granular, and liquid forms. These can be mixed into potting soil, sprinkled onto roots during transplanting, banded over larger seed in the furrow, or coated onto bulk seed for field planting. What counts is physical proximity between the mycorrhizal inoculant and the plant roots. Biostimulants are available to help break spores from dormancy and sustain the fungus as it finds its plant host.[3]

A granular-based inoculum allows for the full range of arbuscular-type fungi to be put in play. The size of spores determines which species can make the grade in liquid or powder formulations. All "ecto spores" are fine enough to be incorporated in micronized products, but some "endo spores" are larger and would be destroyed by grinding. These range in size from 40 μm to over 600 μm across arbuscular mycorrhizal species.[4] Viable root fragments are about 110 μm in length, about the average size of spores in the Glomeraceae family. Powder and liquid formulations are typically passed through a #50 screen, which allows particles

Larger seeds, such as beans, can be inoculated with a gentle peppering of spore powder directly in the furrow.

under 300 micrometers in size to pass.[5] All but one of the high-performing species listed just ahead (see page 94) qualify.[6]

Soluble formulations make larger-scale applications possible. Be aware that mycorrhizal propagules settle out quickly in water and must be continuously agitated to remain in suspension. This makes drip irrigation systems suspect . . . considerable lengths of trunk line and tubing will not be delivering spores on the far end, regardless what sales reps promise. Liquid formulations are best delivered by seeding equipment that treats seed directly before it's put in the soil. Choose a rainy evening if making a surface application . . . as daylight is detrimental to a life-form that otherwise does its thing underground.[7] AM spores are formed in the soil and engage with roots in the soil, so ideally need to settle in at least an inch below grade whatever the method of application.

Cost depends entirely on the concentration of propagules in the product and the efficacy of the application. Time for some big numbers.[8] Crumbly, sweet-smelling topsoil contains six billion propagules to the acre, spread evenly throughout the top 6 inches (15 cm) of soil, whereas this can drop to six hundred million propagules to the acre in average topsoil. Counts in depleted topsoil come in at three hundred thousand propagules to the acre and often considerably less. The quasi relevance of this is that a given root can find its fungal soul mate a mere eighth of an inch (3 mm) away to as much as fifty times that given equal grid distribution of spores. You get the picture . . . isolated roots are not necessarily going to make any connection at all.

Meanwhile, a single pound of granular product typically contains sixty thousand propagules of endomycorrhizal fungi.[9] Assuming each spore could be placed equidistantly results in a relatively substantial stretch to the next feeder root over (forty times the optimum), which is why manufacturers specify using as much as 60 pounds of inoculum per acre.[10,11] Price that out at typical retail charges and you will find the cash register ringing up at nearly a thousand bucks a pop. (Definitely ask your supplier about bulk pricing.) Needless to say, if a company doesn't provide propagule counts for its product, take the money and run. Those of you working with ectomycorrhizal inoculum for forest restoration purposes have it far easier, as 1 pound of these products easily contains 110 million to as many as 2 billion spores. Direct inoculation of root systems — seedlings and saplings — takes the

guesswork out of placement and thus requires far less inoculum on a per-acre basis.

One thing to keep in mind about the "down and dirty" attributes of both AM (endo) and EM (ecto) fungi featured in the inoculum hit parade is that certain good traits are universal. Like improving soil structure. Or snagging phosphorus. Real mycorrhizal fungi do what needs to be done.

The mix of species in commercial inoculum may be relatively limited, yet it's well chosen. The primary criteria are adaptability to varied soil types and a wide array of benefits. Plants exhibit comparatively little host specificity, which shouldn't surprise us, given the large number of plant species to fungal species involved in symbiotic partnerships. Soil pH is the biggest selective factor, followed by soil texture (sand, clay, silt, peat, and saline) and availability of organic matter. The fungi utilized in commercial inoculum blends tend to have wide tolerance ranges. *Glomus etunicatum* is at its best in the acid range, scavenging for nutrients less apt to be released. Another stalwart, *Glomus intraradices*, happily attaches to nearly any plant, whatever the climate, and produces abundant spores to boot. Some species get going like gangbusters in cool spring soils, and others bide their time until the soil warms up. Some help seedlings get established, while other species take the lead in more stabilized ecosystems.

An inoculum mix with multiple species doesn't just up the odds that one or two will serve in the life of an annual planting. Or go on to say goodbye once woody perennials have established in healthy fashion with native fungi. Chances are there's a niche wanting to be filled even in the most diverse landscape. Introducing regional strains can enhance the functionality of a local strain of the same arbuscular species. I like the "shotgun approach" all the more as I become better

Classifying *Glomus*

An interesting thing about species lingo is that old assumptions fall to the wayside as we learn more about genetic pathways. The Glomeraceae family has recently been reclassified into more distinct genera, thereby changing long-standing names in what was once mostly *Glomus*. Take *Glomus intraradices*, for instance. Officially it's now *Rhizophagus intraradices*. Additionally, many species reported in the literature as *G. fasciculatum* appear now to be synonymous with *G. intraradices*. Which has been deemed *R. irregularis* in yet other genomic circles. Such debates are in full swing. Meanwhile a guy could go loony! Inoculum names in the first edition of this book will match the earlier names, which are still being used by most suppliers.

familiar with the players. I also have no doubt that mycorrhizal networks possess a diplomatic touch, negotiating for outcomes that will carry forward. Ebb and flow works that way.

Others prefer a more economical option, especially turf managers and growers with significant acreage. We could call this the "silver bullet approach," but that's way overplaying the hand. Single-species inoculum features *G. intraradices* in the starring role, often in combination with select helper bacteria. One of those is *Azospirillum brasiliensis*, a nitrogen-fixer for cereal grains and turf grasses, recently introduced by Reforestation Technologies International as part of its biological product line. Strains of this AM fungus abound, with thirty molecular sequences noted to date. You may encounter companies promoting geographical isolates as especially promising (desert strains in particular) but just know that *G. intraradices* performs best in soils that are nutrient-rich, regardless of any hard sell. Jump-starting annual crops and making grass on golf course fairways

more resilient has merit . . . as long as we recognize the difference between singular inoculum and more enduring solutions.

Caveat emptor means 'let the buyer beware' in Latin. Those who market biology can assume that most potential customers don't really have a handle on the facts. Finding companies one can trust makes a huge difference. Showering praise on a single species of mycorrhizal fungi is fine as long as it's clear that's what you're getting. Using spore counts to pitch quality, unfortunately, is where the wool can be pulled over people's eyes. Spores of *G. intraradices* are on the smaller end of the range for endomycorrhizal species, and thus far more spores can be packed into a pound of inoculum, especially when produced without soil. As many as six times more per gram. Sounds good, yet there's nevertheless a ploy in how such numbers are promoted. Greater diversity of species brings into play larger spores and highly infective root fragments. Better value won't necessarily be reflected by eye-catching spore counts.

ENDO HIT PARADE

GLOMUS AGGREGATUM
(NOW *RHIZOPHAGUS AGGREGATUS*)

- Functions well in sandy soils
- Assertive root colonization at seedling stage
- Tolerant of high fertility levels
- Improves soil aeration
- Enhances fruit tree vigor

GLOMUS INTRARADICES
(NOW *RHIZOPHAGUS INTRARADICES*)

- Highly effective in a wide range of conditions
- Quick colonization of seedlings
- Phosphorus go-getter
- Increases crop yields
- Improves performance of turf grasses and nursery stock
- Provides disease resistance against fusarium
- Mediates heavy metal toxicity

GLOMUS MOSSEAE
(NOW *FUNNELIFORMIS MOSSEAE*)

- Stabilizes soil aggregates
- Prevalent in disturbed soils
- Increases nitrogen and phosphorus uptake
- Reduces plant uptake of arsenic
- Moderates drought stress
- Increases flowering and fruiting
- Improves performance of woody ornamentals

GLOMUS ETUNICATUM
(NOW *CLAROIDEOGLOMUS ETUNICATUM*)

- Good scavenger of low-fertility soils with high acidity
- Consistently beneficial in agricultural soils
- Very effective in reducing drought stress
- Cofactor influence on protein synthesis

The actual manufacturers of commercial inoculum are relatively few. Mycorrhizal Applications in Oregon crafts many of the reputable brand names featuring customized spore blends. Production from this one company alone now tops 1 million pounds annually. One burning question (perhaps) is how exactly are propagules produced in large quantities?

Open pot culture introduces a chosen host plant to a desired arbuscular fungal species in a contained space. One growing season later, spores and root pieces alike can be separated from soil by washing the soil through very fine sieves, concentrated by centrifugation in a dense medium such as a sugar solution. A number of species are still done by means of wet-sieving and decanting (*Gigaspora*, for instance, which requires six months or more to sporulate). Growing host plants in actual soil to produce inoculum is relatively low-tech, and therefore more accessible to local entrepreneurs. In vivo production by definition results in a less concentrated product due to soil media being part and parcel of the inoculum mix.

The aeroponic technique of spore production involves suspending host plants so that exposed

- Increases aromatic plant production of essential oils
- Improves plant establishment
- Helpful in strip mine reclamation

GLOMUS CLARUM
(NOW *RHIZOPHAGUS CLARUS*)

- Well adapted to a wide variety of plants and soil conditions
- Increases phosphorus uptake
- Impressive outreach of mycelium
- Mediates heavy metal toxicity
- Promotes salt tolerance
- Improves nitrogen fixation activity
- Increases grain and legume crop yields

PARAGLOMUS BRASILIANUM

- Thrives in rich organic soils
- Stimulates root enzyme activity
- Protects against heavy metal toxicity
- Enhances soil remediation efforts

GLOMUS DESERTICOLA
(NOW *SEPTOGLOMUS DESERTICOLA*)

- Common in semiarid and arid conditions
- Improves yields in alkaline soils
- Very effective in reducing drought stress
- Promotes salt tolerance
- Delivers phosphorus from nearby rock
- Increases nitrogen fixation activity

GIGASPORA MARGARITA

- Indigenous across many ecosystems
- Forms extensive external mycelia
- Phosphorus go-getter
- Improves production of tropical and subtropical fruits

Credit for compiling an earlier rendition of this information is due Dr. Mike Amaranthus of Mycorrhizal Applications.

roots can be continuously misted with a nutrient solution in a closed chamber below. One interesting twist here comes by way of *transformed roots*, which produce pure strains of starter inoculum in vitro.[12] This has proven useful in producing clean root propagules and AM spores in a shorter time period. Roots from host plants are generously colonized after ninety days in the aptly named bioreactor. The resulting inoculum is more concentrated, but tends to have a shorter shelf life. So far only a limited number of species show compatibility with this growing environment, with *Glomus intraradices* leading the way.[13]

Quite a few ectomycorrhizal fungi are simply collected from the forest. Those species that fruit belowground, including truffle-like species of the genus *Rhizopogon*, lend themselves to wild harvesting. Each fruiting structure is essentially a pouch of spores awaiting further application. Other EM fungi can be cultured on an agar growth medium in the laboratory. Portions of mushroom sporocarps or mycorrhizal short roots will go full cycle in producing more fungal biomass for use as inoculum.

All spores lose viability over time. Germination will not initiate in proper storage, yet youthful zest never lasts forever. The "infectivity" of AM

ECTO HIT PARADE

Rhizopogon spp. could be called the "false truffle" genus. These fungi play a key role in the recovery of conifer forests following disturbance. Species typically propagated in the trade are *R. villosulus*, *R. luteolus*, *R. amylopogon*, *R. rubescens*, and *R. fulvigleba*. Found on every continent except Antarctica. Benefits include:

- Promotes soil structure (like all mycorrhizae do!)
- Tolerant of cold soil temperatures
- Adapted to a broad pH range
- High levels of enzyme activity benefiting micronutrient acquisition
- Utilizes organic matter as nitrogen source
- Protects seedlings against moisture stress

Scleroderma spp. sporulate for the most part underground as tough-skinned puffballs. Widespread across an array of diverse habitats, from hardwoods to conifers. Often found in short grass and mossy areas near trees. Top-performing species are *S. cepa* and *S. citrinum*. Benefits include:

- Rapid early growth of inoculated tree species
- Increases nitrogen and phosphorus uptake
- Stimulates feeder root production
- Rhizomorph structure improves moisture distribution
- Amelioration of heavy metal toxicity
- Improves restoration of degraded soils

Laccaria spp. are a northern temperate mushroom genus with edible qualities, notably on the hot side. Affiliations vary across birch, beech, oak, pine, and spruce groves. The two species most recommended are *L. laccata* and *L. bicolor*. Benefits include:

- Improves survival of inoculated tree species
- Increases nitrogen and phosphorus uptake
- Stimulates feeder root production
- Protects roots from pathogens
- Tolerant of high fertility levels
- Decreases drought stress

Pisolithus tinctorius is widespread across an array of diverse habitats, particularly sandy soils. Host plants include conifers and oaks. Known as "dyemaker's puffball" for imparting a reddish-brownish color to wool. Benefits include:

- Rapid early growth of inoculated tree species
- Stimulates feeder root production
- Tolerant of high soil temperatures
- Amelioration of heavy metal toxicity
- Inhibits soil pathogen growth in rhizosphere
- Valuable in disturbed environments and acid soils

Credit for compiling an earlier rendition of this information is due Dr. Mike Amaranthus of Mycorrhizal Applications.

propagules registers at 90–95 percent out of the starting gate. Fresh granular product includes mycelia, which might last a few weeks at best . . . and accordingly no one reckons this aspect. Root fragments containing hyphae carry over for six months or more, thanks in part to carbohydrate reserves in the root tissue. These propagules are more successful than spores, unit for unit, but truth is, by the time commercial inoculum gets bagged, shipped, and out to the field . . . the better part of viability may have already declined the invitation.

Manufacturers will tell you to expect an extended shelf life of up to three years for spore inoculum stored in dry conditions at room temperature. Doubling the application rate in subsequent years will make up in part for the loss in viability of the root fragment portion. Better yet to figure two years max and then renew the subscription.[14] Ectomycorrhizal spores, while extremely numerous, germinate in lower proportions (10–20 percent, give or take, in presence of host roots) but retain viability for ten years or more.[15] Providing a short chill period to any inoculum product prior to use (two weeks in the fridge at 39°F/4°C) can nudge germination rates up noticeably.[16] Mycorrhizal inoculum is susceptible to environmental stress, like any living material — so don't allow the container of inoculum to sit in the sun or expose it to a repeated cycle of freezing and thawing.

Homegrown Methods

You can find hyphal connection for your trees and shrubs in the wild. Or you can grow arbuscular-type propagules to apply in the garden from local populations. The first approach will bring out your *Inner Ent* while the second will get the neighbors talking.[17] Time for some ecto and endo improvisation.

Several considerations come into play when gathering "ent soil" to use to inoculate woodsy plantings at home. Indigenous sources of ectomycorrhizal fungi are as nearby as a healthy stand of trees in undisturbed ground. Think *deciduous* if your plantings will shed leaves; think *coniferous* if your plantings are evergreen. Chances are the right sort of generalist species will be found in the ground beneath a vigorously growing specimen. Nor does it take much soil to introduce these hyphal fragments to the ground back home, whether at the time of planting or tucked near roots beneath a recently mulched tree.[18] The upper 4 inches (10 cm) of soil contains nearly three times the mycelia that will be found three times deeper down. One scoop of soil duff per tree or shrub will do, taken from the top layer of soil from beneath the gifting plants. If fine roots can be seen where you dig, all the better. Going about this with a sense of gratitude honors all the beings involved.

A tad of research into the fungal nature of what you are planting will reveal certain exceptions to an all-out forest approach. The roots of fruit trees, for instance, are considered to be entirely arbuscular.[19] The requirements shift to an endomycorrhizal focus . . . and there's no better place to go than a healthy apple tree in a wild setting for appropriate inoculum. Similarly, a forest-edge ecosystem with wild berries, goldenrod, meadowsweet, and the like will feature a diverse mix of suitable fungi if you can't locate an uncultivated fruit tree. You'll likely be gathering spores as well as hyphal fragments on an endo quest, particularly if this is being done in the late-summer and fall months. Odds are that what are good for apple trees will be good for peach trees, given the broad applicability of arbuscular fungi.

Growing "grass in bags" can indeed set off neighborhood curiosity. (No doubt only adding to your flourishing reputation!) Yet it's within a small volume of soil, with the right choice of plant, that mycorrhizal spores can proliferate and be readily available for harvest. Odds are that healthy soil from your home environs contains twenty or more species of symbiotic fungi, all adapted to that soil and its seasonal rhythms.

Cultivating arbuscular-type fungi is relatively easy. The root systems of host plants such as bahiagrass or annual rye form affiliations with a wide range of mycorrhizal species. Start seeds of either in soil contained within a fabricated bag of heavy-duty poly. These *myco sacks* need to be approximately 16 inches (40 cm) high, stapled on the fold to measure about the same across, and set upon ground cleared of vegetation or on landscape fabric. The soil within needs to be crumbly topsoil taken from where wild grasses and herbs grew with vigor and vim the season before. Amend this with vermiculite or perlite or even coarse sand (one part in five) to improve aeration, but definitely do not add any enhanced NPK fertilizer.[20] Work with forest-edge soil if you have woody perennial plantings in mind. Either source brings native species of mycorrhizal fungi into play. Conversely, you could use sterile potting soil and a few teaspoons of purchased inoculum to multiply a commercial blend . . . but that totally misses the point of selecting for site-based fungi. The addition of vermicompost (produced by earthworms) to the rooting medium has been shown to significantly increase spore production.[21] The planted grasses in these sacks need to grow for a period of several months, during which time arbuscular fungi develop multiple associations with the host plant. After that time, cut the grass back radically and stop all watering. Or simply let frost do the job in northern zones. Growth cessation sends a message to the fungi to sporulate like mad in preparation for the next growing season. Harvest the roots of

Growing inoculated bahiagrass or annual ryegrass in myco sacks will prove extremely cost-effective for folks desiring to work with local mycorrhizal species.

the host plant and all clinging soil in late fall, or wait till earliest spring. Chop this into finer pieces and set aside in a cool place until needed.

David Douds developed this system at the Rodale Institute in Pennsylvania.[22] This will never fly on a megascale but works for two types of operations. Vegetable and herb growers who produce their own seedlings can mix the inoculum into a spring potting mix. Or earnest homesteaders can incorporate the inoculum by hand directly into a planting furrow or tree hole. The results will initially be hidden from the eye until healthy growth unveils that fungi adapted

Holy Hydroponic Fungi, Batman!

Marijuana growers employ mycorrhizal fungi in hydroponic growing systems. No soil is involved. Time to think through basic assumptions, whether under the influence of tetrahydrocannabinol or not.

Admittedly this is controversial environs for plants and fungi alike. The reliance on soluble ion uptake is strong, to say the least. Lack of legitimate organic matter eliminates normal prospects for absorption of partially built nutrition. *Houston, we may have a problem.* Yet organic-minded growers insist that symbiosis dynamics do not waver even in an artificial growing environment. And when you consider the oceanic origins of plant-fungi dualism . . . maybe the odd guys out have a valid point.

Arbuscular mycelium can be looked upon as a secondary root system for plants. Those fine fungal hairs are much finer than root hairs, and so have a bigger surface area by which to absorb nutrients. Keep in mind that over 1 mile (1.6 km) of hyphal filaments can exist in one thimble of soil. Why not the same in a misty realm? Mycorrhizal enzymes are integral to chelating essential minerals to a more bioavailable form. Research has also revealed that arbuscular mycorrhizae promote root branching and the energetic feeding capability of feeder root tips. Expanded root systems in turn lead to an increase in bud yield. Biological protection from root zone disease puts the icing on this cake.

Hydroponic inoculum like the highly recommended Great White and Orca brands contain a slew of helper bacteria as well as mycorrhizal species. Ganja jamming, mon.

---

to local soil conditions thrive once more. Viable spores do indeed abide when this is done right, on the order of one hundred million propagules in a single enclosure, according to the meticulous Mr. Douds. *There's gold in them thar bags, boys.*

Carbon Pathways

Further investment in mycorrhizal fungi will be made with carbon currency. The continuum nature of soil carbon pools begins with root exudates and microorganisms. Plants die; animals die; microbes die. The organic matter that becomes soil is truly layer upon layer of life reinvested for the next round. Organic residues in turn get broken down into smaller and smaller particulates, going from recognizable origins to the infinitesimal. Something happens at this point whereby soil carbon becomes more resistant to continued microbial degradation. What's long been called *humification* results in the dark, long-lasting aspect of the soil known as *humus*. Hold on to that thought.

Microorganisms form a considerable percentage of soil carbon as a result of divvying up root exudates. Yet it's fair to say everything is in flux at this stage of the process. Respiration by living organisms (including plant roots) sends a good portion of carbon back to the atmosphere in the form of carbon dioxide. Additional CO_2 is recirculated as a result of oxidation of organic matter. These loops of the carbon cycle mesh beautifully with photosynthesis as plants return carbon back to the soil through root exudates and plant residues. Prospects for a net deposit of carbon staying in the soil are determined primarily by mycorrhizal fungi and helper bacteria.

Those microorganisms in the soil food web that tuck away carbon are crucial to this planet being a habitable place. The glomalin left behind by AM hyphae represents 30 percent of the structural carbon found in soil. This glycoprotein would be considered a "nonhumic biomolecule" in the soil carbon pool, given its penchant to resist further degradation for a decade or more. This is the sticky glue that binds soil particles together. Soil aggregation (see "The Glomalin Connection," page 70) in turn provides space for smaller bits of microbial organic matter to collect and be subject to biological alchemy.

Those last two words have been deliberately chosen to reflect an ongoing debate between soil biologists and soil chemists. Time to bring growers into the fray! We all have some notion of what the term *humus* represents in soils, that dark matter that crumbles in our hands and smells so sweet. The alpha and the omega that beckons green plants to grow. It's as real as my soul and yet even now almost as mystical. So why do we need to reexamine basic assumptions of how organic matter becomes humus? It all comes down to the surprisingly undefined nature of *humic substances*.

Definition suggests clarity. Soil organic matter derived from plant material, animal remains, and manure is all "stuff" you can initially see. The residues of such contain carbon and other simple compounds, which bacteria and fungi consume and subsequently release. These sources of *labile carbon* provide the energy required for the assimilation and mineralization processes to repeat endlessly, regardless of whether they are in coarse form or broken down into more digestible particulates. This is the functional carbon that can quickly be consumed by soil organisms because it's bioavailable (in the form of oils, sugars, and alcohols) and physically accessible to microbes.

Smaller organic bits get retained within soil aggregates held together by the carbon sugars channeled to soil via the hyphae of mycorrhizal fungi in association with actively growing green plants. Bacteria involved at this stage of the digestive process play a major role as a protein source. Still other organic bits will be bound tightly to silicate clays or held in an organo-mineral form. A line has been crossed. These darker, colloidal substances all consist of *resistant carbon* that gives durability to soils. The humic portion of soil organic matter results from the humification process, and requires nitrogen, phosphorus, sulfur, and several catalysts (including iron and aluminum), in addition to fungal-mediated carbon. Humus cannot form to any significant extent without mycorrhizal plants on the scene.[23] Just as importantly, hungry critters no longer have easy access to further degrade what remains.

The only thing that's certain is that without humification of some sort, all carbon will oxidize back to the atmosphere. Organic carbon moves between pools: some short-lived, others enduring.

A controversy began back in the 1930s, when future Nobel laureate Selman Waksman pointed

Table 5.1. Soil Carbon Pools

Soil Carbon Pools	Description	Nutrient Role	Persistence	Rating
Living organisms	Root exudates, bacteria, actinomycetes, fungi, protozoa, nematodes, arthropods, earthworms	Primary source of carbon and energy	Days-years	Labile
Fresh residues	Identifiable plant, animal, and microbe remains and metabolites	Energy, carbon, and nutrients	Months-years	Labile
Particulate organic matter	Simple compounds (sugars, starches, amino acids)	Nutrient assimilation	Months-decades	Labile
Humus	Nonhumic biomolecules (glycoproteins, waxes, lignins); organo-mineral complexes	Nutrient retention	Decades-centuries	Resistant
Recalcitrant	Charcoal	Rechargeable exchange sites	Centuries-millennia	Stable

out that techniques intended to reveal the chemical and physical nature of humic substances — whatever such might be — resulted in extracts that were in truth products of the extracting media and not true constituents of SOM. He wrote: "One may feel justified in abandoning without reservation the whole nomenclature of *humic acids* . . . beginning with the *humins* . . . and ending with . . . *fulvic acid*. These labels designate, not specific compounds, but merely certain preparations which may have been obtained by specific procedures."[24] Apparently this realization got lost in the whorls of war. A renowned scientist clearly stated that humic substances obtained by alkaline extraction (using sodium hydroxide) are chemically and physically different from the organic materials that actually occur in soil.

This shatters widely held notions of what makes for abiding soil fertility. Almost everyone these days speaks in terms of humic acid and fulvic acid as the chemical building blocks of rich soil. That somehow a transmutable synthesis occurs with elemental organic matter to create hefty molecular structures with persistent qualities. Certainly all sorts of biological products on the market proclaim the glory of these two organic acids. Still, does this "point of order" actually change anything?

Not really. Humus is too tangible and (let's be honest) beloved a term to not exist . . . even if molecular analysis now confirms we've been adhering to a false synthesis.[25] Similarly, let's take humification at face value: Soil carbon gets stashed away with or without our knowing how. That's nonnegotiable. Where we should give ground is in getting specific about humic substances. The compounds known as humic acid and fulvic acid result from a human-synthesized process. Neither exists naturally. Both are, in essence, man-made chemicals. (Certified organic adherents will have to come to grips with that hot potato!) Therefore, by the powers vested in me by the Mighty Mycelium, may I suggest, henceforth and forevermore, we call these *the humic mysteries* and be done with this just dispute? I'm only kidding, of course, as what's really to be found in humus already stands revealed.

101

The handful of soil in the upper left is the result of years of excessive tillage. The next handful of soil reflects what results with that same ground after spreading compost and growing winter annuals. The darkest soil of all comes from the pasture on the other side of the garden gate . . . where nobody screwed things up in the first place! Photo by Gowan Batist.

Soil organic matter in relatively stable forms consists of the remnants of bacteria and fungi — those smaller and smaller bits — either as fragments of cell walls or as individual molecules. Even recognizable, in fact, as seen by means of a scanning electron microscope. These minute particulates attach to mineral surfaces within soil aggregates and between the crystalline plates of silicate clays. All will decompose further . . . it's just that these are relatively secure places to hang out for decades on end.

The grower perspective on carbon breaks out according to the color of our deeds. The same soil carbon pools are involved, the same fungi and bacteria — only the language takes on the beauty of earth tones across the seasons. Less jargon, more art, essentially. Management choices influence soil organic matter by either increasing levels of stable carbon or drawing down reserves. The right sorts of doing that ultimately build healthy soils for the duration underscore good farming.

Green carbon provides nutrients for the next crop. Shallowly incorporated organic matter in the form of cover crops and plant litter can maintain the condition of upper layers of soil for productivity. Farmers with a cover crop mind-set throughout the year are making great strides in restoring soil health while reducing the need for fertilizer inputs. Fresh organic matter that is rapidly mineralized doesn't so much increase humus content of soils (0.1 percent per year[26]), but other processes are at work. Deep root systems in a cover crop mix, aerate, and break up the soil while nourishing a wide assortment of fauna and flora. No-till and reduced-tillage practices up the humus ante considerably by leaving much of the green cover in the form of surface residues. Grazing livestock between plantings or spreading fresh manure to launch biomass production on poor soils falls under the green carbon banner as well.

Brown carbon feeds fungi. Manure-based composts supply primarily short-lived humus that supports primarily bacteria, which is all to the good, as bacterial metabolites in turn supplement fungal ecosystems. The humus portion of compost contains nitrogen, phosphorus, and sulfur, which replenish nutrients being removed from the soil by cropping. Regular applications of compost teas will introduce more humus-oriented organisms into soil to create more durable carbon pools. Use of ramial chipped wood or fungal compost follows the forest trail that leads to long-lasting fertility. Systems involve more woodsy perennials and trees, from orchards and community food forests to agroforestry. The fungal crescendo continues to build as soils get disturbed less and less.

———

Black carbon is the soil's reserve. Carbon sequestration produces the real thing. Humus. Both green carbon and brown carbon certainly steer organic matters this way. The journey from decomposition to humification will always be a continuum of assimilation, respiration, oxidation, and eventual stability in tucked-away places. Soil aggregates have a direct correlation to healthy plant metabolism in that the hyphae of mycorrhizal fungi tap in here to more complex forms of nutrition. We have reached grand slam territory.

Significant gains come as roots reach ever-deeper down. Soil ecologist Christine Jones in Australia has undertaken a multiyear study to compare two different soil realities.[27] The one is under conventional grazing, intermittent cropping, and standard-practice fertilizer management; the other is through cropping and grazing management practices designed to maximize photosynthetic capacity. Over the last ten years, the amount of resistant soil carbon (that is, the humic fraction) in the second profile has doubled in the 10–20 cm increment, tripled in the 20–30 cm increment, and quadrupled in the 30–40 cm increment. In future years, it is anticipated that the most rapid sequestration of stable soil carbon will take place in the 40–50 cm increment, then later still in the 50–60 cm increment. That equates to 2 feet (60 cm) of fertile, carbon-rich topsoil continuing to build downward into the subsoil as a result of nondisturbance practices.

All this begins with the sun's energy captured in photosynthesis and channeled from aboveground to belowground as "liquid carbon" (those root exudates loaded with sugars) to fuel microbial action. Minerals are released, a portion of which enable rapid humification further and further down in the soil. The remaining minerals are returned to plant leaves, facilitating an elevated rate of photosynthesis, which in turn means even more carbon gets channeled to the soil, enabling the dissolution of even more minerals. Conventionally fertilized pastures don't achieve anything like this, as the "plant-microbe bridge" is dysfunctional in those systems.

We can further abet the black carbon stage when reclaiming poorly managed soils. Carbonized wood in the form of biochar (see "Biochar," page 108) adds a rechargeable port of call for mycorrhizal fungi to tap into exchangeable nutrients. Humate deposits are mined sources of long-lost fertility, in a sense, found in the transition from peat to lignite, which in turn becomes coal. Humates act foremost as an organic chelator in making trace minerals more available to plants.[28] The granular form will last ninety or more days in the soil exposed to microbes excited to "have at it" once this carbon-rich deposit no longer stands protected by the millennia. Nutrient-charged biochar and humates help fungal systems get established.

Getting carbon in the ground through the interaction of plants and biology results in good soil structure and water-holding capacity. When soil loses carbon, it becomes hard and compacted. The differences in infiltration and moisture retention between high- and low-carbon soils are dramatic. Actively aggregating soils have good tilth. Carbon pathways provide the framework to get where we need to be.

Biological Equipment

The Amish gauge the usefulness of a tool by its impact on community values. Forgo whatever disagreements that brings up for you and just listen a minute. Keeping the phone out on the telephone pole means family dinner conversations are not interrupted. Disallowing rubber tires certainly keeps the economy local. Having no television shields one's heart from the violence of a world gone mad. Selective use of technology involves judgment of larger considerations. Too few people ponder such things. Yet the Amish have shown the tenacity to question the powerful forces of modernity in order to preserve their traditional way of life.

What has that to do with mycorrhizal fungi and healthy plants? Perhaps just about everything.

Tools help humans accomplish the work we deem important. The right equipment can help us build humus, apply trace mineral cofactors, and sustainably manage woodlots. The wrong equipment can destroy mycorrhizal networks, efficiently apply glyphosate across vast acreage of genetically modified crops, and indeed decimate community values.

The lens through which we determine our world matters. Aligning with fungal imperatives — whether in farming, orcharding, or forestry — should be the primary consideration. The principle to *first, do no harm* is not merely about medical ethics. It is about you and me and whatever work we do in this God-given life.

Getting back to a more hands-on scale is part of the rich future we must face. How

Hugelkultur

Burying woody debris creates substantial pockets of organic matter where fungi will prosper for years. White rot species abound in this lignin-rich biomass. Ectomycorrhizal fungi affiliated with living tree roots work right alongside their forebears in extracting nitrogen and other nutrients from trees gone by. The roots of herbaceous plants tap into arbuscular mycorrhizal networks with hyphal access to this same carbon pool. Fruiting

shrubs and trees explore the bounds between species in overlapping mycorrhizal networks. Most pertinently, this collective fungal action promotes long-term humus building.

The German term *hugelkultur* loosely translated means 'hill culture,' whereby rotting logs, whole branches, and forest floor debris are buried beneath soil and other organic matter in a mound. The proper implementation of this technique involves piling woody matter up at or perhaps just slightly below grade, biggest wood first, then covering all to create a raised embankment. Stack logs and larger branches together like a jigsaw puzzle to

This PTO-driven wood chipper provides ramial chipped wood for building lasting soil fertility in Lost Nation Orchard.

I love to use a scythe in my mountainside orchard . . . yet a tractor-mounted sickle bar will always suit down a longer row. It's the biological purpose underlying the mowing that makes the tool. The broadfork, the wheel hoe, and the walking tractor serve us well in our gardens, but I can certainly appreciate wise use of a subsoil shank and spader to transform compacted soil across rolling terrain. That diesel-fueled wood chipper of mine finds its biological groove in the form of ramial chipped wood for building lasting soil fertility. A good ol' pulp hook helps bring in firewood for the kitchen cookstove.

Industrial economics aren't going to be worth one damn dime on a dead planet. The underlying purpose of "biological equipment" is to help sequester carbon. Technology that allows humans to destroy ecosystems must be put aside. It may indeed be appropriate to reconsider the wheel.

limit empty voids. Crisscrossing smaller branches and twigs between upper layers will keep the finished mound from eroding off the sides. Initially, the wood core will take up half or more of the volume in such a pile, resulting in a dramatic-looking upthrust in the landscape. Serious mounds reach as much as 6 feet (2 m) high with relatively steep sides unless covering material is especially abundant. Moldy hay works well on the sides to help keep the core moist. This all mellows in the course of several years as the wood decomposes, eventually looking more like a raised bed on flatter ground or a well-placed terrace on mildly sloping terrain.

The microclimates created on each side of the mound suggest certain planting strategies. Annual vegetables such as squash grown on the heights take advantage of rapidly released nutrients from compost and other mulches. Bacterial heat will be evident that first season and sometimes two (from decomposition of the stacked organic matter), which can help bring in a late crop. Voles can be a concern until things settle, so wait a couple years before rotating in root crops such as potatoes. Fruit trees situated along the edges of the mound benefit from the immediate windbreak. Solar gain comes into play on the south side of these earthworks,

HUMUS
COVER

COARSE
ORGANIC
MATTER

WOODY
CORE

The buried wood in a hugelkultur mound becomes a rich source of fungal doings for many years to come. Orientation of the mounds produces an array of microclimates for growing different crops.

enhanced all the more by any rocks placed to hold raised ground in place. Similarly, gourmet greens can flourish on the north side by virtue of additional shade. Smaller fruiting shrubs and broadleaf herbs do well on the slopes, the roots of which contribute to stabilizing the embankment.

Hugelkultur increases water and nutrient retention as the woody debris breaks down. Herein lies the magic. Saprotrophic fungi at work on the lignins and cellulose in the wood free up vast amounts of particulate organic matter. Arbuscular mycorrhizal fungi specialize in taking those fertile bits, adding some carbon glue (glomalin), and creating soil aggregates. The plants involved are the source of that carbon, through photosynthesis, as provided by root exudates. This fungal banking scheme requires all three components in order to become a megadepository of soil aggregation.[29]

Think rich, dark, long-lasting humus . . . which is what will be found a decade or more down the road if digging deep into a former mound. Turnover of mycorrhizal biomass into the soil carbon pool is thought to be rapid, and in this regard hugelkultur pushes the usual margins.

Buried wood breaks down faster than wood on the surface. Water stored in larger hunks of decomposing wood accelerates microbial decay rates by stabilizing temperature and preventing desiccation during the dry season. Moist conditions attract burrowing and tunneling mammals and invertebrates that improve aeration. Conditions for all sorts of fungi are ideal, which in turn increases the retention and cycling of nutrients within the mound ecosystem. The carbon-to-nitrogen ratio in woody debris piles can run high, suggesting nitrogen will get tied up initially, but let's think about that. Smaller branches in these piles are "ramial unchipped wood" and thus not so suspect. Rotting logs, on the other hand, do contain significant levels of locked-up carbon, but also come with assorted rot fungi already at work. Anticipated nitrogen shock by microorganisms has been mellowed in transit, as it were, with the biggest wood buried far from developing root systems at the start. Softwood has a place in hugelkultur as well, for recalcitrant lignins have been placed in a white rot haven abetted by consistent conditions as can only be found down under. This advice shifts to *pee on your pile* if using all fresh logs, as urine can supply additional nitrogen during the initial breakdown phase.[30]

Riffs on this idea are many . . . but first let's emphasize the availability of waste wood as the principal reason for going about hugelkultur. Importing woody debris from afar makes little sense, whereas land-clearing projects, ready access to fallen trees, and significant coppicing opportunities are site resources demanding to be utilized. Masanobu Fukuoka, the Japanese grower-philosopher, buried grubbed-out (unwanted) trees to create drainage basins in his rocky subsoil as well as to provide reserves of organic matter and nutrients. Deliberate "rubble drains" of buried wood were fashioned to channel water enriched with nutrients released by decay, often directly to favored fruit trees. Using hugelkultur in conjunction with swales on contour on mildly sloping terrain may make sense if vast amounts of brush need to be accommodated.[31] Brush piles neatly oriented (with butt ends stacked tight on the downhill side) and positioned at abrupt changes in terrain can be deliberately torched to add a biochar edge to these proceedings. Stockpiled stumps and loose earth along the upper half of such a pile can be used to squelch a fully consuming fire, leaving behind carbonized ends and brushier nutrition. This then gets

buried completely, thereby smoothing over rugged terrain in the process.[32] Trees planted on the cusp of such woodsy micropockets take off like all get-out.

Hugelkultur mounds flatten out over time, regardless of scale. Raised beds can be replenished with compost and mulches to maintain a reasonable finish height above grade. Stonework adds a nice touch to give annual planting beds definition. Succession proceeds into long-term perennial production in the case of piles left undisturbed in a landscape or orchard setting. Spreading this soil investment elsewhere rarely suits until original plantings have run their course . . . and by then you may not even be able to tell.

Biochar

The vast porous surface area of biochar particles allows nutrients that would otherwise be leached through the soil to be retained in a carbonized

This raised bed was formed atop a shallow trench of decomposing wood, then covered with upturned sod and compost. The flat stones hold edges in place and provide solar gain for heat-loving plants. Courtesy of Dhomestead.

matrix and held at root level until needed. Assorted bacteria and mycorrhizal fungi utilize this same crystalline structure as a haven from the usual feeding frenzy of microbe consuming microbe. Biochar functions like a soil aggregate in a sense, one that comes packed with extreme nutrient density.

What's known in our times as *biochar* came into existence long before humans walked this planet. Lightning strikes caused fires that subsequently converted a portion of plant biomass to "char" . . . and this blackened form of soil carbon proved resistant to further biological decay. Approximately 1–3 percent of organic matter consumed by an open fire converts to stable char, with the rest going the way of mineral ash and greenhouse gases. The fertile soils known as *terra preta* in the Amazon basin in South America resulted from an insightful approach to slash-and-burn agriculture in preparing fields. Pre-Columbian farmers buried smoldering embers in trenches to better lock in the carbon content. Similarly, the recurrent burning of grasslands on the Great Plains by the Native Americans (made to improve hunting odds by congregating game) inadvertently stored significant amounts of stable carbon in prairie soils.[33] Biochar can be viewed as the modern adaptation of such practices.

Scientists in the Agricultural Research Service have shown that quality biochar has the potential to improve soil carbon sequestration by fourfold.[34] Those blackened particles become the basis for enhanced macroaggregate formation, which in turn represents greater glomalin production. This explains why biochar helps to lessen compaction and improve aeration in clayey soils. On the other side of the spectrum, sandier soils with a low cation exchange capacity benefit from the nutrient retention and water storage aspects of biochar application.

Mycorrhizal fungi are eager whatever the reason. Biochar serves as a springboard to further mycorrhizal abundance and functioning by assorted mechanisms, with establishment of a refuge from fungal grazers high on the list. Root tips draw near to biochar particles, but it's only the finer hyphae of symbiosis partners that can access the tiny microsites within the "carbon sponge" to redeem nutrients.

Loading the biochar matrix with minerals helps to get things started. Putting pristine charcoal directly into the soil will have very little immediate effect beyond water retention. Carbon molecules catch and retain exchangeable nutrients continuously, and though the initial stockpiling in a soil environs can take some time . . . once under way, the recharging ability of biochar will keep fertility to the fore for decades to come. This process can be hurried along by saturating biochar with nutrients before it is added to soil. All this takes is to soak the charred particles in compost tea, infused worm castings, liquid fish, or even urine for a day or two.

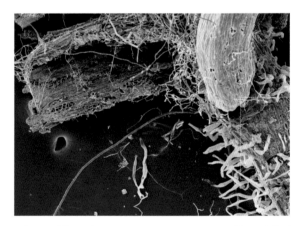

Fine mycorrhizal hyphae extend from the bean root on the right-hand side of this revealing image into the "carbon sponge" of a biochar particle, where exchangeable nutrients are constantly renewed. Photo by Johannes Lehman, Cornell University.

Biochar is extremely adsorptive, variably charged, and electrically conductive — all of which facilitates the chemical reactions that can make the minerals in rock dust available as well.

Pyrolysis of biomass in the absence of oxygen to drive off volatile gases leaves behind blackened chunks of organic matter. The properties of biochar are greatly affected by the type of feedstock being heated and the efficiency of the process. Herbaceous sources such as switchgrass and digested fiber ("cow pies") have higher nutrient content initially, but it's the woody char made from branch prunings and coppiced hardwoods that has greater staying power. An industrial pyrolysis process that captures energy and carbon alike is by far the most efficient. This technology relies on capturing the off-gases from thermal decomposition of woody matter or grasses to produce heat or electricity, while at the same time retaining the ember portion as a useful by-product rather than burning the biomass completely to ash.[35] Homegrown approaches range from a covered pit to burn barrels to a collier's earth kiln. So-called lump charcoal for gourmet grilling (no chemical additives, please) can be purchased locally and then pulverized into smaller particles to facilitate aggregate formation.

Fungal objectives can be achieved by mixing a couple cups of biochar into a planting hole for a tree. Providing this "rechargeable battery" in the vicinity of a developing tree root system in conjunction with a developing hyphal system launches robust growth all around. Improving soil aeration of high-clay soils with biochar requires some means of incorporation to distribute the carbonized organic matter deeper down. A broadfork accomplishes this far more gently than tillage if going about this in garden beds or further out from established trees. Biochar can be top-dressed and subsequently scratched into the surface of sandy soils to increase cation exchange capacity. Finer particles will migrate over time into the entire root zone due to the filtration nature of lighter soils. A respectable amount of biochar in either case would be spread up to a half inch deep . . . which amounts to approximately a cubic foot to cover 30 square feet of biointensive ground. (Think a generous centimeter deep if going about this metrically, requiring approximately 10 liters per square meter on a bulk basis.)

One mycorrhizal twist should be noted for those generating homegrown spores for use in the greenhouse and market garden. Including pelletized biochar in the soil sack for on-farm production of AM fungus inoculum has proven to be fruitful.[36] Hyphae emerging from inoculated biochar particles up the infectivity ante that much again.

Phosphorus Addendum

Mycorrhizal fungi are the principal means plants have for obtaining phosphorus . . . the middle letter in *NPK* as represented by those three omnipresent numbers on a bag of fertilizer. Unfortunately, some serious confusion exists for those who don't grasp that slow-release sources of minerals are far different from the chemical big bang theory. Here are the facts.

Roots occupy only a small part of the soil volume. Immobile nutrients such as phosphorus do not disperse readily, moving only about one-tenth of an inch (2.5 mm) in the course of a growing season. This results in a nutrient depletion zone around roots left to their own devices. Even fibrous-rooted plants in the surface layer of soil where root density is greatest occupy a mere 3 percent of total soil volume. Arbuscular fungi improve the odds of phosphorus assimilation a thousandfold by filling that soil volume with hyphae.

Consider now the first days of a bare-root tree bereft of all fungal friends. Adequate levels of phosphorus are required for generation of new roots. A biologically minded grower inoculates said root system to increase phosphorus uptake by fungal means. An NPK-minded grower floods the planting hole with soluble phosphorus by means of a chemical fertilizer application. Some people do both.

Those high levels of soluble phosphorus at the time of inoculation will not harm the mycorrhizal spores but will halt the germination process. Furthermore, roots that find satisfactory levels of "phosphate soda" aren't as inclined to enter into a carbon deal with fungal partners. Some consultants will tell you that mycorrhizae are "lazy" as a result . . . but this is not the case from the fungal partner's perspective by any means.[37] Conversely, if nearby phosphorus levels are low, roots will put energy into excreting organic acids to increase the availability of relatively insoluble compounds. This in turn provides an excellent growth substrate for soil microorganisms to get the ball rolling, but at a cost. Producing those organic acids requires that fledgling roots consume as much as half of all the carbon allocated to the root system. Less carbon to trade puts the kibosh to red-hot mycorrhizal deals. *Everything in moderation* is starting to look pretty good.

Let's put some soil test numbers to this. Phosphorus readings come in three sizes, if you will — Mehlich, Bray, and Olsen — depending upon the extract used to pull this essential element from a soil sample.[38] I prefer to evaluate phosphorus (and potassium) levels from a middle-of-the-road perspective, so what follows are pure Bray-1 ranges expressed in parts per million.[39] Mycorrhizal fungi spores stay dormant while levels of phosphorus are above 70 ppm but awaken just fine in the 20–50 ppm range. The question is will you be monitoring the math in each and every planting hole? Of course not!

Let's talk about "scattered phosphorus" instead. A trace amount of phosphorus is always at hand, all the more accessible in organic matter, yet bound to run lean in a small space.[40] The roots extend further, with mycorrhizae in the lead. Keep in mind that this is a *developing fungal root system* in ground that has been recently disturbed. Encountering a grain of slow-release rock phosphate here and there totally suits a mycelium finding its way. Deals made in the underground economy shift accordingly as the energy costs of mycorrhizal associations are less for roots than the alternative of producing organic acids. Feeder roots gladly take a breather.

Bringing an establishing root system up to speed requires a touch of phosphorus in the right form. Avoid the overkill of synthetic fertilizers such as Triple Super Phosphate. Soluble phosphorus dissolves in water and so is too readily available. Even fresh manure will oversupply soluble phosphorus. Rock phosphate is okay because it's insoluble, organically approved, and slow-release. Total phosphate content typically exceeds 30 percent, which only becomes available in stages over a five-year period or longer. Promised returns on this sort of fungal banking look good, given the ability of mycorrhizal hyphae to tap into phosphate-rich rock for nutrients as needed.

The upshot? Do as I say when planting a tree.

CHAPTER 6

Practical Nondisturbance Techniques

Building and managing soils from the top down is a fungal maxim. What we do in our gardens, landscapes, orchards, woodlands, and farms needs to take into account the mycelial connections that make for healthy soil. The insights shared in the wide-ranging essays of this chapter are intended to prompt, to poke, and to push us all into doing better by the fungi. Take to heart what suits the work you are doing to be a good steward in your corner of the world. And don't think these techniques are limited to the section defined: One plants a tree in the orchard as well as in the pasture being converted to an agroforest as well as in a diverse landscape. Keeping fungal consciousness to the fore helps keep the choices we face more attuned with how life on this planet evolved. And then we can marvel at the fact that we get to be in this green healthy world, too.

Garden

Getting one's hands in the earth to build good soil turns out to be equally good for the gardener. Antidepressant microbes in soil trigger an immune response in our bodies, which results in the production of higher levels of serotonin.[1] Let's take that jolly state of mind along as we go out to root around in the earth.

The scale at which we approach growing food and medicinal plants determines the tools at hand to get the work done. What's done in a family garden isn't radically different from the techniques employed in a community market garden. Increasing acreage often brings with it the need for larger equipment, but even then fungal principles remain the same.

The whole of green promise starts with seed. What an opportunity to make spore connections for appropriate plants from the get-go. Prepping ground for planting follows from there. Choice of cover crops, the how of incorporating organic matter, and sensible cultivation all tie to what our tools and muscles are capable of achieving. Healthy crop succession is absolutely a fungal affair, despite what you may read elsewhere about vegetable gardening being bacterially dominated. That ol' human tendency to separate keeps popping up . . . but we will not stray from building humus while at the same time reaping nutrient-dense harvests. Interspersing perennial plantings into production plots can take many forms, all of which will prove to be fungal launching pads for annual neighbors. Preemptively spraying for health (as opposed to spraying for symptoms) totally changes the paradigm when it comes to pest and disease challenges. The holistic core recipe utilizes fatty acids, nutrients, and biology to abet healthy plant metabolism as well as healthy soil dynamics. Fostering natural systems comes back to us in foods filled with minerals, vitamins, and those important phytonutrients that help our bodies be strong.

The integrated tapestry we weave in our gardens is very much like a mycelial network. No tomato grows alone. Snap pea roots touch base with strawberry runners. Chamomile blossoms spread good cheer. Arbuscular fungi tend to the needs of many while we gardeners merely need to keep the organic matter coming. You bet a guy can be plenty happy out in the midst of all this.

SEEDLING STAGE

That pinch of seed in your palm is one act away from the union that has made plants strong for millennia. Even good compost lacks the biological connection that will most launch germinating seeds onward to healthy plant metabolism. Sterile potting soils are completely useless. An accompanying pinch of endomycorrhizal inoculum makes all the difference.

Greenhouse seedlings are typically given high levels of water and nutrients to get off to a convincing start. Yet lack of mycorrhizal affiliation discourages plants from producing extensive root systems, which makes transplanting less stressful. Adapting to the real world outside is not merely a question of hardening off to the breezes that blow. The application of mycorrhizal inoculants at germination makes for stronger plants that are protected from root pathogens and have the ability to thrive from day one in the field. That inevitable downtime to reestablish growth thus becomes unnecessary.

The easiest way to achieve this isn't by dusting every seed one by one but rather by introducing spore inoculum into the potting soil. A well-crafted mixture of loam, peat, sand, and minerals should include living compost for its contribution of beneficial bacteria. Sprinkle "myco dust" across the surface and then stir to mix dormant spores throughout the potting soil. You can fill germination flats with this biologically prepped potting mix or make soil blocks for larger seeds like watermelon. Each embryonic root (known as the *radicle*) will quickly come across its "spore mate" given the relatively small volume of soil allotted each seed. Brassicas as a group do not participate in mycorrhizal symbiosis, so you can forgo inoculating potting soil for cabbages and broccoli if seeding significant quantities of those crops. (Otherwise don't worry about a few homeless spores.) Watering seeds with diluted seaweed extract introduces kelp hormones, which help awaken fungal beginnings.[2] Germinating hyphae find the emerging radicle within those first weeks, and then proper arbuscular affiliation begins.

Seedlings like tomatoes and peppers that require "potting up" should be all set from here on in. Fledgling root systems will carry the mycorrhizal connection onward into ever-larger containers, so there's no need to keep adding spore inoculum to additional batches of soil mix for seedlings already inoculated.

Open-pollinated varieties often form surer associations with AM fungi than do more recent hybrids. Newer varieties that have been developed to work well with synthetic fertilization are simply less inclined to take up the fungal banner. Plant breeders reinforce such dependencies by weaving conventional assumptions into trials without a second thought. This doesn't rule out working with hybrids but certainly gives yet another reason to get involved with saving your own seed. Fungal prospects with the seed of genetically modified varieties are no better. GMO corn, which expresses the insecticidal soil bacterium Bt, has verifiable negative impacts on AM fungi.[3] No surprise there when one considers the source.

Fungal inoculation makes sense for field crops as well. A return based on ongoing plant health and increased yields is worthy of the minimal investment in inoculum required. Having spores close at hand to seed in the furrow ensures that symbiosis establishes sooner rather than later. Potato eyes, onion sets, and garlic cloves can be dusted in the tub carried out for planting. Larger seeds such as beans and peas, which are spaced relatively close (an inch or two apart), can be dusted directly in the furrow with a light touch, then covered. Slightly moistening seed will help mycorrhizal inoculum adhere if you prefer to inoculate seed all together. Corn is better done this way given the greater spacing recommended for maize.

Spores require a couple of weeks to form an affiliation with sprouting seeds, and even then the hyphae will be invisible to our eyes. Only home-grown inoculum containing root fragments can up the pace of colonization in the greenhouse.

Northern gardeners have an unexpected edge when it comes to inoculating corn seed. Bare-naked kernels can rot in cold, wet ground unless treated with fungicide. Using such pink-dyed seeds will not exactly be a healthy choice for mycorrhizal fungi.[4] The alternative is to wait to plant organic corn seed outside once the ground has significantly warmed. There's a short enough growing season as it is in Zone 4 . . . so starting corn in prepared soil in plug trays in a warm greenhouse two to three weeks ahead of transplanting time has its advantages. Mycorrhizae synchronize with the emerging corn radicle, helping to hold soil around the root when the day to transplant arrives. Crows turn their beaks up at corn seedlings that are already several inches tall. Rows are completely filled out as no seed rots. What may seem like extra work — dealing with rooted transplants rather than some quickly dibbled seed — turns out to be pretty slick.

The onion family introduces another mycorrhizal strategy. Alliums in general deal with frost better than most early-spring crops. The growth cycle of many onion, shallot, and garlic varieties keys to day length. The sooner these crops can be planted, the more green growth energy will be available come the summer solstice (marking the longest day of the year) to put into the developing bulb below. Gardeners who plan a warmer-season succession alongside inoculated alliums will find that tomato and pepper transplants thrive all the more thanks to an established mycorrhizal network already on hand. Alliums placed along the periphery of a 3-foot-wide (1 m) bed won't be in the way of later crops planted down the middle of that bed. The later vegetables tap into existing mycelia as root systems touch and mycorrhizal hyphae anastomose. The resulting increase in nutrient reach enhances disease resistance in those more tender crops. Leeks extend the same benefits, only without the bulb initiation part of the story line.

Time to do the numbers. David Douds has led research efforts into mycorrhizal affiliation at the Rodale Institute for the last couple of decades. Production trials included plants that were not inoculated and another group that were inoculated with a mix of mycorrhizal fungi. The results over a multiyear experiment showed that the mixture of mycorrhizal fungi increased the yield of marketable-sized peppers up to a maximum of 34 percent over the control. Potatoes showed up to a 50 percent increase over the control. This makes the case that a small amount of mixed AM inoculant can be substituted for a large amount of fertilizer with no loss of yield and considerable cost savings to boot.

INCORPORATING ORGANIC MATTER

Prepping ground for planting through tillage has long been the means of incorporating field stubble and animal manure in the garden. Coarse organic matter disappears below and, oh, what a joy it is to look upon a pristine soil surface ready for the seed du jour. Yet the big problem with this state of affairs is we humans forget to reckon the fungi.

Tilling, plowing, and even double-digging send the soil food web into conniptions. Mycorrhizal networks are literally sliced and diced to pieces. Oxygen introduced throughout the soil strata shifts bacterial diversity in less helpful directions, at least for a time. Any major physical disturbance below the surface of the soil comes with serious trade-offs.

Respecting living soil dynamics came about for me through the development of my very own One-Percent Rule. Somehow, somewhere, I came to understand that every act of tillage costs 1 percent of the organic matter in a given soil. That made things kind of personal, ya know? Especially to a young go-getter being told to increase levels of soil organic matter up, up, up! A guy can really run with an idea like this and, sure enough, I twisted the math completely. My reasoning ran along these lines: "The result of tilling today will be 5 percent soil organic matter going down to 4 percent soil organic matter." That in a mere five boneheaded moves I could absolutely wreck life as we know it. *Well, mister, I'll tell you.*

Holistic perspective goes a long way in breaking the tillage habit. Let's put all the pieces in place and see how we can indeed do better.

***

Early years. Incorporating sod to start a garden or restoring severely damaged soil justifies a degree of upfront compromise on biological principles. Rotary tillers can have a place in the physical conversion of an ecosystem. Scale factors in here, of course, as does the urgency of the plan. Market gardeners have slightly less disruptive options such as spaders and chisel plows to prepare ground. On the other hand, patience rewards permaculture practices such as sheet mulching and forest gardening for those accommodating the long view. Soil compaction must be overcome regardless, and here's where smart application can seem downright counterintuitive.

Breaking up hardpan facilitates deep water absorption and shifts garden soil toward improved tilth. Looser ground obliges seedling emergence and root penetration. Large pore spaces for air infiltration and macroaggregates galore are a mycorrhizal dream come true. But here's the rub. Running equipment over any ground compacts soil, and all the more when soil disturbance deeper down follows ever-larger wheels. Soil compaction has rightfully been called *plowpan* for good reason—a subsurface horizon forms just below the topsoil zone worked above. Despite this, a round or two of "transitional tillage" can be an important means to enable natural processes to resolve this unholy separation. Undoing compaction takes energy . . . and the photosynthetic energy of cover crops is a pretty good way to get it.

Increasing soil organic matter (SOM) with cover crop successions and nearby woodsy mulch will be a lifetime pursuit for any gardener. The primary goal with early cover-cropping plans is twofold. First, get more organic matter in place so that microbes have plenty of food resources. And, second, drive roots downward to bring fungal dynamics ever deeper into the soil. This supercharging of the system will propel years of healthy growing. The takeaway from that math error of mine lies in recognizing that you damn well better have a significant investment in organic matter available to compensate every time when considering tillage.[5]

Roughly incorporating a summertime cover qualifies when the need to seed winter cover soon after makes sense. Smart tillage can be used to reduce rhizome-type grass problems, provided that the preceding buckwheat smother crop(s) add significant biomass in the form of taproots to compensate. Incorporating compost spread a few inches thick qualifies if somehow the crop planned will benefit from a deeper mixing of lime at the same time. Using ramial chipped wood to accomplish a "thin layer transition" toward greater fungal biomass in the fall months (see "Fungality" in "Soil Prep for Fruiting Plants," page 145) requires shallow incorporation to take full advantage of the basidiomycetes' activity at this time.

Ever-increasing soil health makes unmistakable headway. Organic matter levels go up a couple points. Aggregates form. Plant communities thrive. Fungi are in a mellow mood. Tillage becomes less important and indeed all the more harmful.

***

Time passes. Nature builds soil from the top down. No argument there. Perhaps this notion of "incorporating organic matter" needs to be looked at as well. Soil life is extremely adept at *biological tillage* over time. Earthworms literally move mountains in creating humus, while mycorrhizal fungi cement every structural gain with glomalin. The breakdown of organic matter and release of untold nutrients within mostly requires that we humans don't mess with the master plan.

Expanding a garden or simply creating pockets for more perennial plantings can be accomplished by sheet mulching on a larger scale. Sizable sheets of cardboard or piles of newsprint are placed to cover the expansion area. This gets weighted down with either aged manure or compost, along with a sprinkling of nutrient-rich soil amendments like Azomite and kelp meal. Cover the works with straw or ramial chipped wood. This can sit till the next season or even be planted with crops such as winter squash and pumpkins if the manure or compost depth was generous.

Nor do we need to be all that preppy about this. Letting some annual weeds grow in a mulched garden functions as a natural cover crop because, truly, any plant will nourish the soil with sugar exudates that feed microorganisms. Wild greens scattered among crops is not a problem provided that soil-building volunteers are cut down before any seed sets. *Down* as in dropped in place to become surface mulch. Severed roots that quickly die will serve as wicks to improve capillary water flow and aeration. Many "weeds" become nutritious offerings in their own right, from chickweed salad to goosefoot quiche to dandelion root tea. A mixed community of plants brings greater mycorrhizal networking into play as well.

Working with *decomposition tillage* in a garden setting begins with hugelkultur (see page 104). A raised bed formed from buried branches (all those prunings) and fallen tree trunks (dragged from the woods) slowly collapses as the surface layer of compost and rotting hay fills the voids below. No disturbance to organisms occurs, and yet soil strata merge. Similarly, smaller limbs and twigs can be laid down flat as crude mulch, covered or not, creating a fungal haven over the course of several years. Permaculture "chop and drop" is all about that bass improvising on forest floor doctrine.

Maintaining a permanent mulch layer — with whatever combination of straw, autumn leaves, compost, ramial chipped wood, and plant debris — addresses the needs of stockier vegetable plants and fruiting shrubs.

Cruise control. The need for direct seeding in a reduced-till garden brings us to a rift, of sorts. Sticklers can work with *mulch forevermore*, but market gardeners still require open space for traditional row crops and widely planted beds of carrots and baby greens. Here's the skinny on low-impact ways to go about just that.

Garden sections can be nicely primed for seeding with a broadfork and a garden rake. Prepare yourself for one of my favorite spring workouts. A broadfork is essentially a U-bar with several long tines that can be used to efficiently prod open ground. Compacted soils benefit from this treatment, as the tines typically reach as much as a foot deep. I ride the broadfork into the ground with one foot on the crossbar (some people use both) rocking it back and forth as my weight carries the tines down below the surface. Surprisingly little force is needed to accomplish this, especially with heavier renditions of the tool. The actual work comes with moving the broadfork from one spot to the next while pulling back on the handles to add a degree of twisting torque at the finish of each rhythmic step. This is a reverse progression, made by dropping back 4–8 inches (10–20 cm) at a time before the next thrust. Remaining mulch residues and any additional compost you might hope to wriggle into the top inches of soil determines this span. Leaving

The broadfork loosens garden soil in rhythmic fashion without doing in biological connections altogether. Photo by Frank Siteman.

large chunks of earth intact means soil communities are not being torn completely asunder. One pass serves well for larger-seeded crops like beans, whereas two passes side by side provide ample bed width for microgreens. I then use the garden rake to smooth over this rough-and-tumble ground, pulling any extant mulch into what will become the footpaths. Come harvesttime, the broadfork proves useful for prodding up root crops.

Maintaining open ground by means of *occultation* brings an element of virtuosity to market gardening. Use of black plastic mulch throughout the entire growing season has become fairly common, despite being detrimental to a fungal ecosystem.[6] Shorter-term use of heavier silage tarps (8–12 ml thick; UV-stabilized for long life) will have less impact on microbial diversity while effectively creating weed-free beds for planting. Such woven "geotextile cloth" allows both moisture and air to pass through to the soil below, resulting in a far friendlier scenario than what's achieved by sheet plastic left in place for the duration. Jean-Martin Fortier in nearby Quebec utilizes this technique to perfection.[7] Growing beds are covered several weeks before planting. Weed seeds germinate in the warm, moist conditions beneath the tarp, but to no avail, as sunlight cannot penetrate the opaque barrier to support subsequent growth. The tarp comes off and crops go in with minimal disturbance to the soil surface, taking full advantage of the fact that viable weed seed buried any deeper will not become a problem. Larger "occultation mats" can be used to overwinter areas up to 50 by 100 feet (15 by 30 m) for future row plantings.

Tillage radish highlights the fact that select cover crops can prepare ground for next spring's planting even more so. The deep taproots of daikon-like radishes poke holes right through soil compaction and then can be cut off at ground

level (or heavily grazed) in late fall to ensure winterkill. Subsequent root decomposition adds to the breadloaf texture of both topsoil and the upper subsoil. Including oats (which will die back as well) to such a *Brassica* family planting is critical to provide the right sort of root systems for passing along mycorrhizal affiliation.

Enter the walking tractor and an entirely reasonable compromise.

Working the top 2 inches (5 cm) of garden soil to achieve seeding goals is a further concession that soil biology can accept. There will be reasons and seasons to modify the soil surface when growing food crops. This is not "tilling" in the total disturbance sense, but rather permission to be a good grower with a reduced-till mind-set.

Walking tractors are used worldwide for small-scale farming and landscaping. Having a power takeoff (PTO) on its own pair of wheels makes possible a number of functional implements. The flail mowing attachment features a series of hinged blades bolted to a rotating drum that "flails" each cutting edge outward by centrifugal force. Mowed material gets reduced to bits for shallow incorporation into the soil or to be left as fine mulch. A 6-foot-tall cover crop (up to 2 m high) can be reduced to short pieces in a single pass and left evenly distributed across the width of the mower. Root systems and mycelia alike are not disturbed by this complete shearing of the green above. The planting terrain stands roughly revealed for further action.

Another attachment, the power harrow, has multiple sets of tines that rotate off vertical axes (and yes, that's the plural of *axis*). The soil is stirred laterally across a set depth rather than mixed and mashed in true rototiller fashion. Soil layers are not inverted; no compaction profile results. All to the good. A steel-mesh roller in the rear of the power harrow levels and pre-tamps the soil for good seed-to-soil contact. The working depth of the tines can be readily adjusted so as to cultivate only the top couple inches of soil.

These two walking tractor implements make possible a series of biological riffs for dealing with cover crop biomass and prepping a fine seedbed. Our objectives to better care for soil life while building levels of organic matter can be met with thoughtful application. Soil stewards take pride in the fact that a healthy soil will actually keep getting deeper as the years roll by.

Shallow Cultivation

The time-tested way of dealing with weeds in the garden is to prevent errant seeds from germinating in the first place. Mycorrhizal fungi can work with this, as the zone of contention is strictly in that top inch or two of soil.

Mechanical cultivation often increases weed pressure by bringing a new round of weed seeds to the soil surface to germinate. Therefore, let's not work garden ground in such a way that continuously generates new problems. The most effective

The flail mower attachment for the BCS walking tractor quickly converts a stand of winter rye and hairy vetch to fine mulch without disturbing root systems.

The Rinaldi power harrow attachment stirs soil laterally without compacting soil. These features allow market gardeners to prep fine seedbeds and successfully embed cover crop seed with limited soil disturbance. Courtesy of Joel Dufour, Earth Tools.

means to deal with weeds is to strike at the vulnerable sprout stage, when lightly disturbing the soil surface breaks fragile roots and exposes emerging radicles to unrelenting sun. The best tools for this job are assorted shallow hoes.

Onion hoes, oscillating hoes, collinear hoes, and wheel hoes all suit differing applications. I mostly reach for one of the long-handled stirrup hoes with a swiveling, double-sided cutting blade. These well-crafted hand tools cut through weeds just below the soil surface, whether pushing forward or pulling back, making for quick work. Having two sizes of blade widths will be useful in accommodating row spacings that vary by crop. The wheel hoe has a similar oscillating mechanism, only with a much wider blade. These work perfectly to run alongside rows of beans and everything else to keep pathways open. We also use the wheel hoe to maintain garden sections in a deliberate fallow for a few weeks before planting. This quick back-and-forth work creates an old-fashioned "dust mulch" whereby seed bank loading is dissipated for the most part before row crops even go in.

Typically, gardens are cultivated every week to ten days, depending on rainfall patterns and the particular weed pressures faced at a given site. Soil that is overly wet gives young weeds an opportunity to re-root, so wait for dry, sunny days to shallow-hoe. Two to as many as five rounds of hoeing are necessary, especially when early-season weed species are followed by more persistent trouble like galinsoga in the summer months. Keeping blades sharp makes a big difference in terms of efficiency, so keep a keen file handy in the tool shed. All told, you are doing this job right when weeds for the most part are in the sprout stage and can hardly be seen.

Quite a few common weeds, such as redroot pigweed, lamb's-quarters, and wild mustard, are nonmycorrhizal. Others, such as hairy crabgrass, barnyard millet, and green foxtail, are set back by arbuscular-type crops.[8] All produce humongous amounts of seed fast, resulting in "hoeing wars" that can go on for a long, long time. So-called *fungal gardening* will change the dynamics for aggressive weeds in general, as enhanced soil biology shifts the niche advantage to those crops where mycorrhizal affiliation is in place from the get-go. Crops affiliated with mycorrhizae get the phosphorus and other nutrients, leaving the nonmycorrhizal weeds adrift.

Nancy prepares an area for planting with a wheel hoe by knocking back surface weeds without stirring up additional weed seeds from deeper down.

MYCELIUM NEAR AND FAR

Interspersing perennial plantings into annual cropping areas keeps undisturbed ground closer at hand. Such "fungal refuges" can work relatively quickly to reestablish a mycorrhizal network for annual plantings. All the benefits of teamwork follow from there.

Imagine a bed of fall raspberries right down the middle of your vegetable garden. These varieties bear on first-year canes and often don't require trellis support. Mulching with ramial wood chips will keep unwanted plants out of the "cane zone" and help conserve moisture. Fungi will love this. Typically a bed will be kept on the order of 3–4 feet (1 m) wide so berry pickers can reach the fruit in the middle. Extending the mulch alongside the edges of such plantings is optional, as sucker canes will pop up regardless. Shallow tilling for cover crop purposes every other year is one way to knock back the cane undergrowth. Making ready neighboring beds with a broadfork in spring will pop up raspberry roots as well. Soil prep that accomplishes dual goals works for me.

Perennial plantings such as asparagus in the midst of the garden provide a fungal refuge from which mycorrhizal hyphal can spring forth into rows of annual crops nearby.

The garden now stands divided into two blocks on either side of the raspberries. That's helpful for rotating crops and encouraging more consistent cover cropping. No longer are you faced with one big plot but rather you have two manageable units. Adding perennial plantings down the middle of each half breaks this yet again. A bed of asparagus, a bed of medicinal herbs. Echinacea, for one, needs a couple of seasons to develop more sizable roots for harvest. Maybe you go for more small fruits, be it strawberries or a mix of currants and gooseberries, or even a rootstock nursery for a future apple orchard. Another edging advantage now becomes apparent . . . grasses creeping in from outside the garden are no longer right up against more permanent plantings.

Garden designs can gain from a nip-and-tuck mix of kept plants with annual plants. The scale of what's undertaken speaks especially to an "alley cropping" perspective. My weavings in our larger garden plots accommodate one or two or even three tractor widths for when mechanical access suits the work to be done. I might make one pass every other growing season to incorporate rough compost or establish overwintering cover. By all means, whittle this down to the methods you employ in managing your garden space. The permaculture concept of *stacking functions* speaks to additional benefits from intercropping. Perennial plantings will filter the wind, hold soil in place when heavy rains pour down, divert frost pockets, and even provide afternoon shade to nearby greens. The long-term fertility benefit of ramial chipped wood mulch left undisturbed for a few seasons is especially genius.

One of my favorite healing herbs to grow this way is mad-dog skullcap, *Scutellaria lateriflora*. This diminutive member of the mint family starts out as plug transplants from the greenhouse. The young plants are shallow-hoed until well established, at which point a 1-yard-wide (1 m) swath of wood chips can be laid down. Skullcap spreads by means of stolons (root runners) beneath the mulch, thus ensuring a fairly weed-free crop in the second year. Renew the bed with compost or an organic fertilizer blend after that cutting. Semismothering a dense planting with additional wood chips after a second cutting often makes a harvest possible in year three as well. Convert this area back to a cover crop at this point. The ramial chipped wood will be significantly broken down by that time, having been converted to humus by friendly fungi. Richer ground thus becomes the very next harvest to follow three years' worth of formidable medicine.

A trellised (or staked) row of fruit trees especially suits an alley-cropping plan. The dwarfing nature of less vigorous rootstocks works hand in hand with annual vegetable doings. These tree root systems demand less competition in order to be productive. A narrow swath of ramial chipped wood at the base of dwarfing trees keeps the "trunk zone" relatively open.[9] This will prove to be more easily maintained over the years with the addition of taprooted herbs for diversity's sake. Some lovage, some comfrey, some horseradish — typically planted between each trunk where trees are spaced 4–6 feet (up to 2 m) apart. Meanwhile, garden ground lies beyond the outer edges of the mulched swath. This correlates nicely to the spring and fall cycles of tree feeder roots (see "Pulsing the Spring Root Flush," page 149). One year it's potatoes next to the fruit trees. Spuds are hilled and perhaps even mulched with straw. No particular competition for tree feeder roots there. The potato harvest gets followed by an oat cover crop, which will winterkill. Again, perfect for fruit trees starting anew in spring. Year two it's

Ericoid Inoculation

Ericoid mycorrhizal fungi form symbioses with acid-loving plants such as blueberries, huckleberries, and cranberries — all peat-favoring plants with shallow root systems. The fungal skill set required for this job is to break down complex organic molecules in habitats where mineralized forms of nutrients are otherwise not readily available.

Currently, no commercial inoculants for ericoid mycorrhizae exist because these fungi are difficult to isolate and grow too slowly in culture. The indigenous solution is to obtain a small amount of soil and root mass from thriving ericaceous plants established elsewhere. Venture into bogs and across the higher moors if need be . . . though in all likelihood you'll know a nearer place to go on this ericoid quest. Sometimes nursery stock will come with fungal spores by way of peat moss used in the potting mix. Peat containing colonized root debris makes the difference. Effective root colonization with ericoid mycorrhizal fungi was obtained in ten of the thirteen commercial peat products tested in a University of Vermont study.

Blueberry plantings lacking this basic connection are often sluggish. Growers who introduce the fungus several years down the road report exponentially increased growth rates. No surprise there, as up to 80 percent of the nutrient reach of healthy blueberries relies on fungal mycelia.

The farm view on deliberate inoculation of commercial blueberry plantings comes by way of an excellent Sustainable Agriculture Research and Education (SARE) project report (FNE12-770) put together by Ben Waterman of Johnson, Vermont. Check out the detailed instructions for evaluating ericoid colonization in your own plantings, if you want real-time assurance. Apparently "peat serendipity" seems to get 'er done in the end.

beans or onions or peas. Limited soil disturbance is necessary at planting time and then again come midsummer when planting a succession crop or yet another cover. Two things are happening in this swirling dance of green growth and exposed ground. Any shallow cultivation done in establishing the vegetables also happens to serve the needs of dwarf tree root systems. Yet more serious soil disturbance (when necessary) takes place in midsummer between the spring and fall root flush of the orchard planting. Mycorrhizal fungi in turn ebb and flow in sync with the needs of both the fruits and the vegetables.

FATTY ACIDS IN GARDEN SPRAYS

Foliar nutrition backed with arboreal reinforcement gives plants vital *oomph* when it comes to warding off disease organisms. Sprays that include fatty acids build up the defensive cuticle as well, those waxy exudates between epidermal leaf cells and the big wide world.[10] Garden plants offered "holistic homebrew" thrive, whereas those plants put on a fungicide, insecticide, or herbicide regimen end up depleted (to put it mildly).

Boosting green immune function and competitive colonization brings into play an entirely healthy paradigm. The guiding principle to *do no harm* applies as much to soil organisms as to the sentient world aboveground. Bear in mind that a respectable charge of systemic resistance to disease will be induced by mycorrhizal affiliation. We certainly don't want to risk screwing around with that! Spraying herbal remedies, fatty acids, and arboreal organisms are the *coup de grâce* when need merits. What's good for fungi will be good for plants . . . just as . . . what's good for plants will be good for fungi. Caring for whole systems is herbalism at its best.

Holistic spray ingredients typically include seaweed, fish hydrolysate, pure neem oil, and effective microbes in the core recipe. The beat goes on with readily available foodstuffs such as whole milk, fermented plant extracts, and chelated minerals — each chosen for specific crops and the challenges faced. Underlying all this home cooking is the recognition that enhancing plant health rocks at all times.

Principles that apply in the holistic orchard are just as apt for flowers, vegetables, and herbs. One case in point highlights how this should be forevermore.

Late blight fungi were responsible for the Irish potato famine in the mid-nineteenth century. Several years ago this disease came down hard once again on home gardeners tending to tomatoes and potatoes throughout the East. Turns out that big box stores with garden centers were supplied by the same southern greenhouse operation.[11] Late blight could not have chosen a better vector than sought-out tomato transplants. This disease strikes quickly as spores disperse on wind currents and by rain splash. Cooperative extension guidance to organic growers took the form of spray copper, then spray copper again . . . and then pull the damn plants when copper finally proves ineffective. So goes desperate reasoning when it comes to symptoms the world over.

Meanwhile, the tomato and potato crops here on our farm were already braced for the unexpected. Tomato seedlings started in the greenhouse had been inoculated with mycorrhizal fungi from day one. Potatoes received a similar treatment before the seed spuds were hilled over. Mineral levels run high in soils built with compost and cover crops year after year. Last but not least, my own long-standing habit of spraying the garden on those days when doing the same in the orchard ensures that our crops come "lipid-primed" with microbe friends aplenty.

Let's use our inner eye to observe the scene that invariably occurs. A late blight spore lands on a healthy leaf. A competitive crowd gathers to protect its niche. That latest fatty acid feed already has arboreal organisms in an uproar for more. *Gulp*, perhaps. The *Solanum* leaf itself (tomatoes and potatoes are of the same genus) stands primed by means of induced systemic resistance. The terpenoids in neem and the cytokinins in seaweed have tickled the appropriate phytochemical response beforehand to the enzymatic chemistry now being employed by the blight hyphae to obtain plant sustenance. Just as pertinent, the proteins associated with mycorrhizal arbuscules down in the roots are on call to further the plant defense response above. That "poor little late blight bugger" never had a chance against the full monty of healthy plant metabolism.

Table 6.1. Core Holistic Recipe

Ingredient	What's What	Purpose	Backpack Rate	
			4-gallon tank	15-liter tank
Pure neem oil	Terpenoids Azadirachtins Fatty acids	Immune activator Pest deterrent Microbe support	2.5 oz.	75 ml
Fish hydrolysate	Modest levels of NPK Fatty acids Hormones and enzymes	Foliar nutrition Microbe support Biostimulant	10 oz.	300 ml
Seaweed extract	Cytokinin hormones Trace minerals Polysaccharides	Immune activator Enzyme cofactors Frost protection (liquid kelp only)	½ oz. dry 2.5 fluid oz.	14 g dry 75 ml
Effective microbes	Proprietary mix of photo-synthetic bacteria, lactic acid bacteria, and fungal yeasts	Competitive colonization Organic matter catalyst Blight protection	6 oz. w/ English dollop of blackstrap molasses	180 ml w/ metric dollop of blackstrap molasses

Emulsifying Neem Oil

The raw oil pressed from neem seeds must be emulsified in order to mix into water. Pure neem oil will be as thick as butter at temperatures below 60°F (15°C) — that's good for retaining constituent effectiveness in storage, but not the morning you plan to spray. Place a batch-sized container in a warm room (but not directly in sunlight) for a day until the consistency reverts to a homogeneous liquid. Setting "semithawed neem" into a bucket of warm water on cooler mornings may be necessary to fully liquefy the amount needed.

An emulsifying agent can be any biodegradable liquid soap. This must be mixed directly into the neem oil first, on the order of 1 teaspoon of soap for every 2 ounces of neem oil. This mixture will quickly become "greenish opaque" as the large globules of fatty oil divide into smaller globules. Pour this oil/soap blend into a small amount of lukewarm water and stir. Sizable oil globules floating on top at this stage indicate the need for slightly more soap to fully disperse the fatty components of neem. Water with high mineral content may need to be "softened" with citric acid in order for this to work. The completely emulsified mixture may be added to the spray tank and its full volume of cooler water.

Which isn't to say gardeners can forgo common sense about disease cycles and obvious sources of inoculum. We've seen very little late blight over the years as a result of holistic practices. Yet stuff happens. Watch for "hot spots" in your potato and tomato plantings — any plants showing blight symptoms should be immediately pulled and destroyed to prevent further spread of inoculum.[12] Clean up completely after harvest if blight raised its ugly head that season. All leaves, stems, unripe fruits, and even roots can be sent to the landfill in a plastic bag or placed on a burn pile.[13] Late blight especially fancies a living host to survive between seasons . . . therefore be sure to pull volunteer potato plants that sprout up the next season from potentially infected spuds that escape notice at harvest. What you don't need is a launching pad for a disease this devastating right in the midst of it all.

Some sprayer tips for gardeners are in order to keep everybody happy. A backpack sprayer has several advantages over the handheld type, provided you have body size to support as much as 40 pounds (18 kg) of weight. The first can be kept pressurized by means of a pump lever at hip height, whereas the other needs to be set down continually to be pumped. Backpack sprayers typically have a 4-gallon (15 l) capacity — enough to roam freely when applying holistic homebrew to all sorts of plantings. Spray mixes containing fatty acids need to be sloshed as you move about: A little shimmy action with that sprayer on the back will keep the oily portion from settling on the top. You gain an ergonomic edge by positioning the full backpack (with all the fixings) on a rock wall or appropriately high stool so as to slide into the harness easily.

How simple and satisfying *spraying healthy* can be!

COVER CROP ESTABLISHMENT

An important biological caveat will always be to maintain healthy communities of native mycorrhizal fungi in the ground itself. Spore production is good, but colonized root fragments are even better. Add "fungal futures" to the lengthy list of virtues provided by wise cover cropping.

Honoring the maxim of *no bare ground* (excepting short interim periods of deliberate fallow) requires we understand effective means for establishing cover crops. Seed germination in the hands of the clueless doesn't get very far. Let's first review the basics:

> All seeds germinate and establish best with good seed-to-soil contact.
> Smaller seed cannot be covered too deeply, if at all.
> Larger seed can be covered up to several times its diameter.
> Incorporation following broadcast ideally includes an element of tamping.

Variations on these themes apply to different groupings of cover crops. Grasses are more adapted to germinating on the soil surface, which allots some leeway to how oats, barley, and rye can be placed. The radicle sprouting from a grain seed can easily penetrate the surface crust provided moisture levels remain adequate.[14] Small-seeded legumes do well if covered ¼–½ inch (6–12 mm) deep in the seedbed. Red and sweet clovers can be established in earliest spring through *frost seeding* when conditions allow freezing and thawing to "drop" each tiny seed into the soil surface. Large-seeded legumes, such as field pea and vetch, insist on complete seed-to-soil contact. Brassicas readily seed in any cover crop cocktail (see "Cover Crop

Cocktails," page 177) but do not carry mycorrhizal association forward as a sole planting.

So be the rules. Next come the moves.

Working with hand tools in small areas can be quasi effective. Broadcast larger seeds by hand atop cleared ground (where the aboveground portion of crops has been either mowed down or removed to the compost). Use a garden rake to "knock open" the soil so as to get the majority of the seed covered. The stirrup blade of a wheel hoe does a decent job around root stubble, provided you master the back-and-forth flip necessary to throw soil over more seed than not. These methods are not brilliant, but indeed serve in a pinch. Clover and brassica seed can be scattered across the planting area immediately after this rough incorporation; the hope is that rounded seeds will roll into moist folds of earth. A light smattering of compost thrown across the exposed seed aids germination considerably.

Planting into more formidable stubble is a trickier business on a small scale. What's needed is a "dibble on steroids" capable of pushing seed through to the rich earth below. More practical people simply seed their choice of grain heavily (as much as three times more than a farmer drilling seed with big equipment) and try to whack that seed down to moister environs beneath flail-mowed mulch. Calling in the "venerable beast" for situations like these can be forgiven — garden rototillers set to take no more than a hand-width bite will incorporate a mix of cover crop seed with pretty good results. Mycorrhizal setback is minimized, in part because the approximate working depth of the tines will somewhat be kept afloat by surface residues.

The definitive tool for incorporating and tamping in cover crop seed is the power harrow attachment made for the walking tractor. The

Red clover pops up with renewed vigor after mowing the nurse crop of oats. Round Two of soil building continues with no additional effort.

horizontally spinning tines can be set to take seed to a precise depth, and the mesh roller follow-up ensures righteous seed-to-soil contact. A full-width drop seeder can even be mounted above the harrow unit for landscape seeding across open ground.

Market gardeners into extreme minimal impact can plug into farm hack ingenuity.[15] A lot of the work a full-fledged tractor does requires relatively little horsepower. Furthermore, shallow cultivation aimed at incorporating seed is best done at a slow speed. Plans have been freely shared to modify a bicycle into a "culticycle" where human

Spotlight on Sudangrass

The moment I saw "the chart" in *Mycorrhizal Symbiosis*, I knew my favorite cover crop was so much more. Listed were the arbuscular affiliations of *Sorghum sudanense*, more commonly known as sudangrass. Five fungal genera were unequivocally identified by the researchers with the roots of this plant: *Scutellospora, Gigaspora, Acaulospora, Entrophopora,* and *Glomus*. But the beat didn't stop there by any means. All told, fifty species of mycorrhizal fungi have a sure thing going with sudangrass. Fifty species, dude. The prospects for fungal futures never looked better.

Sudangrass and its hybrids — crossed back with forage sorghum — can grow up to 12 feet (4 m) tall, thereby producing more organic matter per acre than other major cover crops. Seed cost is modest to boot. These fast-growing, heat-loving, annual summer grasses can smother weeds, suppress some nematode species, and penetrate compacted subsoil if mowed once.

Soil temperatures reaching 65°F (18°C) are required for good germination . . . which gives sudangrass the green light throughout the United States and southern Canada. Sow at least three months prior to the first hard autumn frost to maximize spore formation of mycorrhizal fungi. Adequate rainfall is always good, but this maizelike grass can withstand drought by going nearly dormant, once established. Sorghum-sudangrass hybrids are often used in rotation with barley to reclaim alkaline soil.

Here's where things become all the more interesting. Mowing or grazing when stalks are a few feet tall encourages *tillering* (basal shoot growth) and a deeper root system. Think "significantly more penetrating roots" in fact, as sudangrass becomes a subsoil aerator in its own right. A single mowing on New York muck soils caused roots to burrow 10–16 inches deep compared with 6–8 inches deep for unmowed plants — an

power propels the cultivating tines. Being able to pedal down a long swath of willing earth can be a game changer in getting cover crops covered with no diesel involved.

This cycle of seed will sprout and subsequently connect with mycorrhizal spores and hyphal fragments left behind by the previous crop. Cover cropping carries symbiosis forward to the next planting. Round and round and round we go.

FUN GUYS DO GARLIC

Getting a fungal leg up in the garlic bed is invaluable. Rot woes will be solved by using clean cloves for seed and insisting on the pathogen protection found in a robust rhizosphere. Garlic bulbs are all the more medicinally potent when a diverse "garlic food web" delivers balanced mineral nutrition. And what fun to unearth a crop

Everyone should have a favorite cover crop! Sudangrass brings a tremendous return on inoculum investment while at the same time bulking up organic matter both aboveground and belowground.

increase of as much as 25 centimeters. Root mass typically increases by five to eight times as a result of this shearing of the green. Vegetative regrowth will be less fibrous as well, which aids immensely in decomposition later. Keep at least 6 inches (15 cm) of stubble to ensure good regrowth and continued weed suppression.

Sudangrass does well as a stand-alone planting, with good smothering action due to its bulk and protective allelopathic compounds. Mix in some fast-germinating buckwheat if weed seed loading from pigweed and the like happens to run exceptionally high. A nitrogen boost can be gained by using sunn hemp (the legume *Crotolaria juncea*) in combination with sudangrass in warmer zones.

Making a field-wide investment of spore inoculum with a sudangrass planting maximizes mycorrhizal response for future crops. Decomposing roots will carry this diversity forward till the next spring, along with a disbursement of fresh spores. There's no better way to get the ball rolling.

surrounded by visible white hyphae reflecting a job well done!

All starts with using fungal-friendly compost in preparing garlic beds for fall planting. This lead-in period typically involves a short fallow following a summer cover crop (such as buckwheat mixed with millet) to lessen weed seed loading where the garlic will be going.[16] Mixing fresh ramial chipped wood into activated compost placed nearby (see "Fungal Compost Pointers," page 154) greatly improves fungal prospects. I spread this woodsy mix of compost a few inches thick across each planned bed about a month before planting the garlic.[17] Seed cloves need to go in before the ground freezes solid, yet have enough time to develop fledgling root systems. Warmer falls have pushed back the planting date for garlic here in northern New England to late October if not early November.

Seeing strands of saprotrophic hyphae while harvesting garlic bulbs reflects earlier preparation of the planting bed with ramial chipped wood. Mycorrhizal hyphae are here as well, only invisible to the naked eye.

Lightly incorporating the compost with passes of the wheel hoe (double-rigged stirrup blades are especially nice for this purpose) the week before will fluff up the beds for inserting cloves one by one in grid fashion.

This hands-on seeding lends itself to mycorrhizal inoculation. I bring selected cloves of each garlic variety out to the field in a shallow tub. The works get dusted with inoculum, and as I reach for more cloves, the spore powder constantly gets on my fingers, which then carry additional inoculum along into the ground. A quick look to get the rooted end down and *plop*, each clove is inserted several inches deep into friable soil just begging for mycelial resurrection.

The first days of spring find garlic tips reaching for the sun. Growing sizable bulbs by midsummer keys to growing sizable tops throughout the spring months. A modest side-dress application of organic fertilizer at the first shallow cultivation (when garlic plants are a few inches tall) adds a charge of nitrogen to boost green growth. Fatty acid sprays contribute nitrogen as well, but the real thrust here focuses on robust soil dynamics.[18] *Feed the fungi, feed the plant* . . . and great garlic will surely follow.

Landscape

Friends who are landscape contractors do amazing work for those who can afford top-notch services. Seeing more diverse plantings, lots of healthy trees, even the occasional estate orchard managed for the "unspeakably wealthy" is all to the good. We need fewer lawns and manicured marigolds in this world. Still, it's a stretch for me to imagine not having my own hands in the earth. Those of us with land can learn a thing or two from how professional gardeners and arborists go about creating and maintaining inspired landscapes in fungal fashion.

PLANTING A TREE

A guy digs a hole and puts the roots facing down, right? This ain't exactly rocket science.

The stories we could tell! The basics of planting a tree are fairly straightforward to a "tree person" who thinks about trunk flare and spreading out lateral roots. Green empathy alone can get you fairly far. Reviewing how to do this right, however, brings out some mycorrhizal nuance no one should overlook.

We begin with the tree itself. Nursery stock ships bare-root, but landscape trees from a local foliage farm come either potted up or with a significant root ball. Let's ignore the apparent differences for now and simply keep things on a human scale — as in younger trees with girths that can be manhandled and root systems that are not yet completely mangled or tied up in knots. Still, a field-grown tree (nicely balled in burlap) or container-grown tree has less than 20 percent of the fine absorbing roots of the same-sized tree in the ground. That alone tells you why orchardists prefer to plant bare-root stock. Regardless, we all do our best to ameliorate transplant shock with an initial soak in a seaweed solution followed by copious amounts of watering in the months ahead.

Water stress imposed by that limited root system will be moderated by rapid root growth. That's accomplished by digging a righteous hole

where new roots can reach out into aerated soil and immediately be at home. Tree root systems are wide-spreading, which indicates that the planting hole should be wider than it is deep. A saucer-shaped hole with sloped sides will encourage initial root regeneration upward and outward toward higher-oxygen soil near the surface. It's important that this early outreach effort by the root system penetrates the site soil rather than circling back on itself in a more can-shaped hole.[19] Any taproot inclinations can come later once the tree has become established in its new location.[20]

How deep? How wide? A respectable hole for bare-root stock typically runs about 16 inches (40 cm) deep and at least 36 inches (90 cm) across. Keep that "saucer shape" in mind, for this is not as much digging as you're thinking. The central crater can be more like 10 inches (25 cm) across

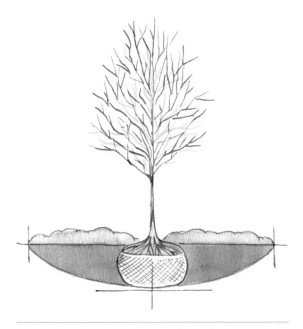

Understanding the dynamics of root outreach by a newly planted tree leads to digging a sloping planting hole that guides roots and mycorrhizal fungi alike toward the surrounding native soil.

at depth, at which point the sides gradually curve upward. Some of that width can be gained by collapsing the rim of the hole when the backfilling begins. I like to place dug soil equally round the circumference of the hole for the amending plans that lie just ahead.

Larger trees require a planting hole that's three times the diameter of the root ball, with the center depth kept a few inches shy of the height of the root ball. The backfill zone thus created has room enough for the root system to rapidly grow four times over, before being slowed by the lower oxygen levels of the site soil. A 2-inch-diameter (5 cm)

Hydrogel formulations of mycorrhizal inoculum add a visual flair to "fungally planting" a bare-root tree.

tree stem with a 2-foot-wide (60 cm) root ball needs a 6-foot-wide (2 m) hole. No more narrow, deep holes! Trunk-girdling roots develop when a tree is planted too deep in the planting hole. You want the root ball to sit on undug soil, allowing for just an inch or two of rise above the surrounding ground.[21]

Time to get fungal. Dipping bare-root systems into "mycorrhizal gel" adds a visual element to these proceedings.[22] The sides of a bigger root ball can simply be dusted in the hole so as to place spores near to roots. One caveat for those planting trees in late fall: Soil temperatures below 50°F (10°C) hinder spore germination, in which case it's better to wait until the soil warms again in spring to place mycorrhizal inoculum in the upper layer of soil near to roots. A lifetime of symbiosis lies ahead for the just-planted tree as a result of investing this teaspoon or so of inoculum.

Yet more considerations come prior to backfilling. Sprinkling mineral amendments atop the sloped sides of the hole and the dug soil at the rim (and even a little way beyond) puts certain nutrients within reach for the establishment phase of the tree. These will be mixed randomly into the returned soil. A couple pounds of Azomite with its full array of trace minerals is a given. Ditto for slow-release rock phosphate (see "Phosphorus Addendum," page 109) to provide scattered pockets of phosphorus until the fungal duff ecosystem comes up to snuff. A modest dusting of alfalfa meal provides a low-key source of nitrogen in weaker soils for modest shoot growth that first year. This last recommendation is entirely for the fungi: Carbon-rich humates will stimulate mycorrhizal hyphal growth, while biochar will provide plenty of pore space for micronutrient uptake. These dark "fungal amendments" can be used generously.

Let's deal with temptation. Some of you are itching to put an impressive charge of soluble

NPK into this planting hole, admit it. Resist this urge. Significantly altering backfill chemistry changes prospects for roots that get overly attuned to the good life. Mycorrhizae will prove slow out of the starting gate if feeder roots can simply lap up what's needed. Nature builds soil from the top down with organic matter, and more on that lies just ahead. Similar advice applies to a mega-charge of compost to modify sandy soil or heavy clay. Sate such desires with a 5-gallon (15 l) bucket of compost (at most) mixed into the backfill soil. Compost adds "critter diversity" in terms of beneficial bacteria and predators (protozoa and nematodes) to the soil food web being developed for each newly planted tree.

Tamping in the roots of a tree is an honor, so go about it that way. Take time to spread laterals. Root balls often will yield a share of winding roots to such outreach.[23] Work soil into voids at the base of a bare-root sapling and then pack things in tight with the palm of your hand. All the while you will be effectively mixing mineral amendments, humates, biochar (if available), and that modest charge of compost throughout the backfill zone. Give all a good soak so as to collapse any air pockets left alongside roots. Finally, if you're truly feeling inspired, grab the broadfork and loosen compacted soil with one pass around the packed hole.

First fungal advantage comes almost immediately as mycorrhizal hyphae lead the way for roots through the backfill into the site soil.

MEANINGFUL MULCH

Building a fungal duff ecosystem for landscape plantings continues with the right mulch. Trees respond quite well to an "expanding radius" approach to building soil. A smattering of compost on the surface creates a rich soil-mulch interface to get things started. Cardboard sheeting can be used out beyond the dug tree circle if sod competition is thick, to then be covered with woodsy mulch as well. Never place mulch right up against the base of the tree. Trunks that stay consistently damp can be subject to bark decay and other management issues. Start mulch about a foot's length out from the trunk, as much as several inches (10 cm) thick over the backfill area and beyond. Alternatively, piling allotted mulch on the downhill side of a tree planted on a slight slope will help retain moisture that first year when watering. The far side can receive a mulch treatment in subsequent years.[24] Likewise, the radius of the mulched area can be enlarged as the tree grows, if that suits the setting.[25] Plantings of perennials or fruiting shrubs out from the trunk are readily made in future seasons.[26]

The ideal mulch for building the fungal component of soils will always be ramial chipped wood. That topic has been thoroughly vetted (see "Ramial Wood Nuance," page 74), so let's move on down the line. Hardwood bark would be an acceptable second choice if your mulch options indeed come by way of the garden center. Turning in a strictly softwood direction will always be dubious. Pine and hemlock barks, often shredded, come laced with natural tannins . . . great for suppressing growth, but please remember, a landscape steward really wants to put the emphasis on actually supporting growth. One fatal step beyond these are the "pleasingly toned bark mulches" colored by toxic dyes. Just don't, and never say you did. Sawdust and wood shavings used for animal bedding should only be considered several years down the road after they've absorbed vast quantities of barnyard urine to counter all that carbon overkill.[27] Fungal mulch should foremost be about soluble lignins for white rots to convert into lasting fertility.

Establishing trees need room to develop root systems. Piling ramial chipped wood deeply on the downhill side of the tree begins the process of developing a fungal duff ecosystem while also helping retain moisture in the root zone.

Using ramial chipped wood as mulch on multiyear plantings is a good way to build lasting soil fertility for whatever comes next.

And why shouldn't we build humus every chance we get? Emulating the natural twig and leaf fall of the deciduous forest is just the ticket for fungi and woodsy plants alike. A mixed chipping containing no more than 20 percent softwood will favor white rots and thus can be spread freshly chipped. That said, granting a coniferous pile of wood chips time to mellow for six months or more gets beyond the allelopathic impact of brown rots.[28] All organic matter is indeed "good stuff," provided we take into account microbial process and intended use. Younger plants are most sensitive to such effects, as they do not have the extensive root systems of established plants. Ramial mulch resists the tendency toward compaction as seen in barks alone. Laying hardwood prunings down on the ground (as small enough pieces to lie flat) works equally well in the long run. The right mulch will always make for ecstatic feeder roots in the fungal duff zone around trees and shrubs.

And yes, let's broaden that perspective once and for all. An artificial division of sorts has been proposed in the gardening world whereby "woodsy plants are fungal" and "annual vegetables and flowers are bacterial" in regard to soil dynamics. The gray area would be perennial vegetables and flowers, which thrive in rich fungal ground given time to get up and running before mulch gets applied.[29] But let's take this further and recall a fundamental fact about mycorrhizal symbiosis: Nearly every plant on this planet benefits from fungal connection. *Nearly every plant.* Most annual vegetables and flowers have to be included here as well . . . which suggests the entire concept of "bacterial gardening" is far too simple a construct.

Truth gets determined at the soil-mulch interface. Concerns about nitrogen lockup have legitimacy. Nitrogen becomes temporarily unavailable to plants whenever soil microbes demand more of this element in order to deal with bulk carbon. That earlier prohibition to avoid sawdust and wood shavings as fresh mulch was so about not binding up nitrogen. Bacterial mulches (such as grass clippings and rotting hay and, yes, bacterial compost) make sense in the vegetable garden, as tomatoes have but one season to fruit and satisfy. Experienced gardeners don't use woodsy mulch materials on open ground around vegetables, primarily because a newly introduced mulch interface may not work out for annual crops in the limited days ahead. Which is far different from saying tomatoes don't like to have their toes tickled by fungi. Most plants do.

So why isn't this mulch interface quandary that pertinent in landscape plantings? Truth is we've been accommodating this dynamic all along. *Give plants room to establish. Cultivate shallowly. Start with aged compost. Place mulch further out for future growth.* Another part of this story line is that permanent mulching will likely be renewed on a regular basis. A transition occurs between the active decomposition zone at the soil surface up through ever-mellowing organic matter to the freshly added mulch on top.[30] Finer roots can choose to venture into that zone or not. Just like on the forest floor.

The one bugaboo to these proceedings comes about when humans apply petroleum by-products to the soil surface. Please do not say *plastic mulch* under any circumstances. Meaningful mulch breaks down. Meaningful mulch contributes nutrients and soil structure. Meaningful mulch welcomes fungal friends. Meaningful mulch creates a soothing place to be a root. Meaningful mulch becomes enduring humus. Plastic films and landscape fabrics do none of the above.

Use of plastic in landscapes as a long-term approach to weed control is a poor move, biologically speaking. Pull up landscape fabric after a number of years and beneath you will find that the soil is dead. Soil that started out crumbly and loose becomes hard and impossible to dig. Organic matter is gone. Rain runs off the surface. Soil life has left in search of greener pastures. All rates as one big soil fail.

Landscape fabric reduces the air reaching the soil and prevents the normal restoration of organic matter. Life depends on air and food. You don't need to be a soil microbiologist to understand that certain basics matter. Plastic in any form is not good for healthy plant metabolism. Those rolls upon rolls of black polyethylene used in vegetable and herb production do not bestow *food as medicine* virtues to crops. Synthetic fabrics can have practical worth in the short term — warming soil in early spring, occultation to reduce weed seed loading — even provisional lockdown of invasive grasses. Yet ultimately we should be about supporting life . . . and nobody does it better than microbes converting lignin-rich organic matter in the form of ramial chipped wood.

OUTRAGEOUS DIVERSITY

Creating a fungal landscape correlates with the design process that takes place above. The *structural layer* becomes the arbuscular mycorrhizae that glue the works together. The *seasonal theme layer* could be provided by the allure of ripe blueberries and the subsequent bright-red foliage of heath-type plants introducing an ericoid mycorrhizae canopy. Which leaves the ectomyorrhizae of landscape trees as the *grounding layer* of choice. Perhaps even some functional mushrooms will pop up!

Health abounds wherever a diverse mycorrhizal network is recognized as an important design

This landscape planting includes a range of native and fruit-bearing shrubs coupled with the beauty of sugar maple. The woodland fungal canopy has been well served. Photo by Michael Nadeau.

The plants in the ground layer of this rain garden are *Packera aurea*, *Allium cernuum*, and *Deschampsia cespitosa*. The taller seasonal theme plants mingling with the ground layer are *Iris versicolor*, *Vernonia lettermannii* 'Iron Butterfly', and *Eryngium yuccifolium*. The root diversity belowground abets a resilient fungal network in turn. Photo by Claudia West.

parameter. A landscape that thrives in its own right will have plants tapped into multiple fungal systems. Consider the mycorrhizal perspective next time you look out on your favorite view. Native grasses interspersed with clover have the nitrogen-phosphorus dynamic well in hand. Passage plants located between groupings of trees make possible tree-to-tree communication and nutrient support. Bridge trees such as rowan (*Sorbus aucuparia*) bring AM and EM fungi together to potentially share resources between otherwise isolated networks. Balanced nutrition mediated by fungi is the first line of defense against disease running amok, just as is the case in the holistic orchard. The insect prospectus is where things get especially interesting.

All sorts of beneficial insects keep pest numbers toned down to a reasonable rumble. Having this sort of beneficial diversity on hand requires providing untold flowering habitat for adult parasitic wasps, ladybugs, lacewings, and hover flies to flourish. Adult beneficials accordingly lay their eggs nearby to where these nectar-providing plants are found. Voracious larvae in turn feed on problematic insects. The specific plant choices that make this possible demonstrate how natural systems maintain healthy equilibrium. A must-have book about just such plants is *Farming with Native Beneficial Insects* put together by Eric Lee-Mäder and the other pollinator ecologists of the Xerces Society. Nine different ecosystems across North America are examined in detail, with lists and lists of wildflowers and grasses to weave into every landscape, whether food-producing or not. Integrated design projects include hedgerows, field borders (meadows), and conservation buffers. You'll learn which insects are to be celebrated and start to see these very actors when strolling about your place . . . all due to providing the right habitat.

Weaving such diversity into stunning art comes next. An actual *layered landscape* involves thinking about seasonal glory and textural patterns tied to a functional layer that brings resilience to an ecosystem. Broad concepts, I know, yet the hands-on landscape design as featured in *Planting in a Post-Wild World* by Thomas Rainer and Claudia West looks to be a great way to get started. The nicely structured progression presented in this book has me thinking about our farmscape in a new light. Seriously. If a guy like me can achieve a *Monet moment* at this corner of the orchard path or over there with birches and ferns and purple astilbe perfectly timed . . . then why not? Vitality designed into urban landscapes keeps everyone uplifted by the greenery. A little bit of plant investigation and artful innovation . . . and Nature lost becomes Nature regained.

Taking full advantage of "light space" and "root space" involves more practical plant matters. Plantings can often accommodate more companions than might be apparent to the untrained eye. Some root systems run deep and narrow; others are fibrous and spreading. Tree foliage need not impede fruiting shrubs as long as some direct time in the sun comes along for the berries. Airflow between layers helps limit disease potential. Well-designed fruit guilds stack multiple purposes into plantings. Comfrey in the tree dripline serves as a living mulch plant, falling over when top-heavy, making calcium-rich soil for feeder roots. Daffodils nearer the trunk drive voles elsewhere. Siberian pea shrub can be coppiced occasionally for woodsy mulch while all along its roots fix nitrogen within reach of the fungal network of the fruit tree. Lovage and marguerite flowers bring on an array of beneficials. Lemon balm soothes the soul and provides nectar to bumbles. No one root system crowds out the others.

1. Asian pear; 2. buffalo berry; 3. comfrey; 4. lovage;
5. horseradish; 6. pea shrub; 7. daffodil; 8. marguerite;
9. lemon balm; 10. wood chip ground cover.

This polyculture of fruit trees, berry plants, and taprooted herbs shows how light space above and root space below can be fully utilized in a diverse landscape planting. *Illustration by Elayne Sears.*

There are many ways to go about diversity, no question, but one theme I hope many will pick up on is edible landscaping. Why not grow food in the very places we breathe in such beauty? Nor do the crops necessarily need to be just for humans. Watching birds and wildlife partake of berries and seeds is just as much a part of the fun as what we get to put into our own baskets. Not getting enough? Plant more! Small trees and shrubs are often underused in landscapes. Add some elderberries, dogwoods, hawthorns, and aronia berries to the mix. Your heart will swell in the wonder of it all.

TREES FOR THE DURATION

Trees outside the forest face multiple challenges, from soil compaction and limited renewal of organic matter to paving to the nth degree. How roots and fungi manage is beyond comprehension . . . and yet trees indeed do grace city streets and stand tall in heavily trafficked parks. Tree health can best be renewed by making up for lost opportunities.

Many arborists are big on *mycofracking*, which is essentially the application of fungal spores deeper down. Injection tools use compressed nitrogen gas to apply mycorrhizal inoculum, biostimulants, and often compost tea. The idea here is to kindle fungal activity and thereby increase root activity. Such treatments have been shown to turn the tide for trees in decline. On the other hand, research seems to indicate no gain will be had when inoculating established street trees.[31] The difference lies in each tree's setting. Applied biology does not have the same impact under pavement, whereas open ground benefits from the reduction in soil compaction attributed to such injections.

One common scenario of specimen trees responding well to this treatment occurs in suburban woodlots where much of the land is cleared in the process of construction. A few significant trees are kept and a lawn gets planted. What's forgotten is that these trees once had fungal networking ties that are now irrevocably broken. Furthermore, tidy homeowners are raking up and removing the annual leaf drop. Of course the remaining trees go into decline! Bringing fungal sense to the backyard begins with mycorrhizal renewal.

Similarly, the health of a poorly performing tree or shrub brought into a house lot ecosystem can be improved by loosening the ground under its canopy with a broadfork, especially at the outer edges where most of the fine feeder roots grow close to the surface. Sprinkle mycorrhizal spores into these crevices and cover with a smattering of ramial chipped wood to keep the topsoil moist. The fungi will take things from there.

Urban trees particularly benefit from biochar being distributed throughout the bulk soil at planting. Carbonized chunks of wood become surefire aggregates for ongoing fungal activity in soils where organic matter renewal will always be questionable. Biochar will help with water and nutrient retention, while the outreach of hyphae will keep soil structure more amenable for roots.

PERMACULTURE GREENHOUSE

Yearlong growing beds work best when given a woodsy underpinning to feed fungi for decades to come. The structure itself can incorporate the latest in energy innovation without necessarily going high dollar. A sun pit greenhouse built into an embankment relies on the earth warmth to be found below grade. Attached and freestanding designs with plenty of thermal mass follow from there. Growing a perennial polyculture under polycarbonate glass hinges on establishing a "greenhouse mycorrhizal network" to garner the same plant health connections as outside.

Jerome Osentowski of the Central Rocky Mountain Permaculture Institute in Colorado shares how to create just such an inside landscape in *The Forest Garden Greenhouse*. Extending the growing season in colder climates in permanent fashion certainly allows greens production throughout the year . . . but what happens when you up the ante to create a green oasis capable of supporting fruiting plants verging on the tropical as well as tender culinary and medicinal herbs? The savvy discoveries made by Jerome and his team over the years blazed the way to creating a "climate battery" to keep this growing space reasonably toasty with near net-zero energy input. I'm especially intrigued by the preeminence given to fungi and deep organic soils in setting up this space.

Hugelkultur beds are built atop the buried ventilation pipes that bring excess heat to the subsoil for night storage. Logs and chipped wood alike are covered with autumn leaves, then good compost and native soil (to introduce indigenous microbes). Sheet mulch atop beds takes the form of what's available — straw, spoiled hay, manure, more leaves, and chop-and-drop plants — placed a foot or more deep. Add copious amounts of rock dust and mycelium

Building a "fungal bed" in the permaculture greenhouse includes adding layers of comfrey, coppiced Siberian pea shrub, and other lignin-rich materials from the outdoor forest garden. Courtesy of Jerome Osentowski, *The Forest Garden Greenhouse*.

in the form of forest duff and rotting wood chips as layers build. Soak the works with organisms via compost tea, including mycorrhizal fungal inoculum. Annuals can be grown during the first nine months while soil settles and fungi pick up the pace.

And that's when you get to plant the night-blooming jasmine and figs.

Orchard

My awareness of mycorrhizal fungi began with fruit trees. Turns out that an orchard was the perfect place to observe the connection between plants and microbes that creates lasting soil health.

Apple trees and company straddle an important divide in the fungal realm. Perennial fruit species are thought to be entirely endomycorrhizal, and yet there's this "tree thing" that suggests ectomycorrhizal affiliations can't be that far away. You might come to this juncture through permaculture design or the simple fact that outrageous diversity knocks the socks off monoculture. Root systems belowground share "room in the humus," which leads to further revelations about crop nutrient uptake. The fact that mycorrhizal fungi scratch an innate immune response that reaches up to the leaves correlates perfectly with having learned that fatty acid elicitors, herbal remedies, and competitive biology are the genuine way to thwart disease challenges. All these natural advantages soon had me realizing that *healthy orcharding is really all about fungal stewardship*. And it starts with how we prepare orchard ground for planting to taking full advantage of the pace of decomposition in building a proper fungal duff zone beneath each and every fruit tree.

Sharing all this with home orchardists and community orchardists has been a joy. People kicking around with a few fruit trees get to experience all the wonder of buds unfurling and becoming a ripe peach. *Schluuup*. Local growers with a generous mix of apple cultivars share lore and flavors that neighbors won't find elsewhere. It's so easy to dream about all this, even to order and plant trees . . . but then comes the great disconnect leading to so many important questions. The basics are easy-peasy to explain, such as keeping the base of young trees open to grow a solid branch framework. The enthusiasm of new pruning shears in hand needs to be tempered by a lesson in how green leaves photosynthesize and thus produce root exudates that rev up the growth response. Concepts like "ent soil" and "regarding the trunk more akin to earth" get biological juices going, at which point we're ready to slide into microbe-friendly sprays. This can be a tipping point, reasonably so, as what's sprayed in the conventional realm seems to have defined the act. Bigger challenges come when fruit does not set, mysterious spots appear, and branches die back. Every site is different, and the possibilities can indeed seem overwhelming.

A fledgling orchardist is not unlike a nursery whip: Several years are required to develop those strong branches and that rootedness, just like we are given a span of seasons to grasp the nuance and wonders of orcharding. Stick with it. Tenacity can be a very important quality in life. Or at least that's what the trees and the fungi tell me.

SOIL PREP FOR FRUITING PLANTS

Soil preparation done long before planting day garners a number of advantages for young trees. Orchard beginnings typically require some degree of biological compromise to convert an existing ecosystem, break the compaction layer, and address soil imbalances. A number of overlapping considerations go into evaluating any site, which in turn points to a customized plan that ultimately *needs to go fungal*. Take those last four words to heart, as fostering a robust soil biota that supports healthy growth from the get-go is the mark of doing this right.

———

Structure. We start with the land before us. Is the envisioned orchard going on former pasture already progressing toward a fungal-dominated ecosystem? Or is it a depleted cornfield? Or a rich meadow with untold flowering diversity? Or an inner-city lot with more bricks than actual soil? Even an easily reclaimed lawn isn't ready for prime time without some fungal investment.

The physical traits of the land suggest not only what needs to be done, but also set inherent limits. The option to subsoil, for example, requires that ledge outcroppings be few and far between. Running a deep shank down a field compacted by years of tillage and chemicals makes good sense only if it's doable.[32] Unusually steep ground may call for access terracing, whereas a milder pitch will be better served with water-collecting swales. Wet ground always calls for improved drainage. A raised berm can gain as much as a foot in elevation across moist pockets on gentle slopes. Tile drainage needs to run "toward the plug" (low spot) in the terrain prior to the trees going in.

———

System. The envisioned *orchard system* speaks to the amount of trees to come and the room allotted to each. Tighter spacing between trees suggests working the proposed field area-wide. It's simply less confounding to prepare ground for a high-density planting — where the next course of trellis wires for dwarf trees will be going about as far as you can spit — as one continuous whole in those early stages. Trees on larger rootstock spaced significantly apart, however, will do perfectly fine by sheet mulching at each location. Conserving diversity in an overgrown pasture calls for narrowing the scope of cover cropping to parallel swaths centered over every row. This consideration of the big picture will prove pivotal when deciding the expanse of ground to be prepped.

Time line. Obtaining a preferred mix of fruit varieties on specified rootstock sets a certain pace, regardless of one's inclination to plan ahead or not. Placing a custom order with a commercial nursery requires two years' lead time. Even self-directed grafting requires this amount of lead time. Going from "bud" to "tree" takes what it takes, no matter how much you're champing at the bit. This can be whittled down by half if available inventory lists published the summer before meet your needs for desired cultivars and root vigor. Nursery operations invariably graft extra trees and standing orders do get canceled, so it's always worth checking.

All this suits the prepping time line. Cover crop work begun in spring will result in ground ready

One riff for keeping things more fungal in a high-density orchard planting features taprooted and olfactory herbs in the tree row coupled with shallow cultivation of cover crops in the aisleway. Anything beats the use of herbicides when it comes to mycorrhizal health. Photo by Harry Hoch.

for fall planting in warmer zones and spring planting (in the next calendar year) in northern zones. Generally speaking, that is . . . worn-out soils in need of glyphosate healing will certainly benefit from two full growing seasons of cover cropping.[33] The flip side of this is where things get interesting. Those who want trees "now" often proceed to find trees to plant and deal with fungal conversion after the fact.[34] Trees so rushed often linger a year or two in growth response, and nothing I say here will alter that inherent contradiction.

Transformation. What's actually growing on the land at the onset has bearing. Aggressive rhizome-type grasses, once broken into smaller root pieces, know more about "comeback city" than

Cover cropping a swath for the tree-row-to-be allows a multitude of native plants to remain in place on the edges.

most weeds. Johnsongrass, quackgrass, and knotgrass are all in this category. A basic smother plan in such cases starts with a buckwheat cover crop followed by an interim fallow period followed by a second buckwheat cover crop . . . all prior to initiating soil prep with fungal ends in mind. That plan takes yet more lead time . . . which in retrospect you will further appreciate if you never heed this advice and end up pulling rhizomes from young tree root systems by hand. Bindweed and creeping thistle absolutely must be put on notice. Rampant seed loading in an otherwise bare-looking farm field can pop up as well to later overwhelm young orchard trees. No problem keeping relatively few trees free of annual competition through shallow cultivation, but this becomes a whole different ball game if you've committed a couple acres or more to fruiting intentions.

Lest I be misunderstood, this conversation is not about aggressive or invasive species being "bad" elsewhere in the ecosystem.[35] Certain plants are simply not desirable at the base of young fruit trees or in the midst of a new berry patch! And so wise growers act beforehand to do something about that.

Outright conversion of a young deciduous forest brings mycorrhizal considerations into play. Typically, stumps are wrangled from the ground by heavy equipment, perhaps then used to fill in low spots in the terrain. Overtones of hugelkultur come into play for any fruit or nut trees planted there. Other tree root systems in the vicinity, however, should alternatively be viewed as prime sources of native inoculum. Certain actions on the grower's part make this relevant. Clearing stumps from a row-to-be in order to cover crop does not mean stumps need to be cleared from the future aisleway. Here's a good example where "swath thinking" makes sense in an orchard on vigorous stock. Removed trees can be cut fairly level to the ground to allow mowing in the years ahead. Not only that, sprouts arising from those stumps and subsequently mowed can be viewed as an in-house supply of ramial wood. Leaving living roots nearby means mycorrhizal fungal hyphae are reaching out from nearby. I'll actually let poplar and maple sprouts go a couple seasons to keep this fungal connection intact while orchard saplings tie into indigenous mycorrhizal networks.

Similarly, in a diverse meadow setting (with or without woodsy succession under way) many species of flowering plants already have the beneficial insect scene well in hand. Cover cropping a swath to bulk up on organic matter in the row-to-be allows a multitude of native plants to continue in the aisleway. Why go to the trouble of reestablishing a permanent cover like "orchard grass" when worthy volunteers have been on duty long before you had your bright idea?

———

Tilth. Building humus lies at the core of orchard prep. Cover crops add organic matter, which in turn creates a more crumbly soil structure for tree roots to explore. Structural growth on young fruit trees depends on the ground being kept relatively open for several years in the immediate radius around each tree. The right cover-cropping plan helps reduce the nearest competition, which can be further maintained by shallow cultivation or wood chip mulch.

Some combination of tilling, spading, plowing, and disk harrowing is generally necessary to prepare any site for seeding. Such soil disturbance going in is okay — here's a perfect opportunity to get lime deeper down if soil tests reveal a need to address cation balance.[36] I often wait to address additional fertility shortfalls until trees actually

get planted (see "Planting a Tree," page 131). Supplemental nitrogen can be useful in depleted soil situations to launch a first sowing of cover crops.[37] Once green growth is under way, the goal will be to minimize additional soil disturbance by means other than full-depth tillage. Available equipment and seeding ingenuity come into play as we work to further fungal connections in orchard ground.

Determining which cover crops to plant correlates directly to the tree time line. Growers planting trees in early spring can take advantage of the fact that certain cover crops will winterkill, leaving behind nicely mulched ground for digging. Growers planting trees in late fall likely need to knock down vigorous growth with harrows or a flail mower several weeks beforehand. Should it be early summer when you first break ground, with no aggressive grasses in sight, and trees will be planted two years hence, I'd be suggesting a "glory plan" along these lines:

> sudangrass with mycorrhizal inoculation
> flail-mow by early fall before seed sets
> vetch/rye/pea (roughly incorporated)
>   through that first winter
> flail-mow by late spring before seed sets
> oats/red clover/tillage radish (roughly
>   incorporated)
> scythe high when oats start to head up
> red clover root systems await next spring

Further riffs on cover-cropping strategies are determined by where you grow. Moisture levels need to be sufficient for "roughly incorporated" seeding measures to work. Getting a modicum of soil over seeds can be done with the shallowest tillage, chain-type drag harrows, or even an old bedspring. Not all covers do well in hotter, drier climates — millet and trefoil would be suitable prospects there. Every crop needs a certain amount of growth time to accrue full benefits, especially nitrogen-fixing legumes such as clover and alfalfa. Deeply rooted crops such as rye break up soil compaction so that tree roots in turn will seek greater depth rather than all run laterally.

Fungality. Let's be clear as we turn toward the finish line that my making up words should not be an issue. *Of course fungality!* This is the song that soil life sings for trees. All the emphasis on limited disturbance in process must be met with limited disturbance yet again in actually putting fruit trees and berries in the ground. Tillage at this stage of the game to "level the playing field" would be anathema to the fungal ecosystem we've been working to create.

A quick word about where fruit trees want to grow. Orchard health correlates directly to the ratio of fungal biomass to bacterial biomass found in the living soil. A biological community featuring ten times higher amounts of fungi than that of bacteria turns out to be ideal for apple trees and the like. This ratio describes the soil food web as found on the edge of a forest. Preparing orchard ground through cover cropping and with ramial mulches has been all about creating the same fungal ecosystem in full sunlight.

Fulfilling any cover crop plan with red clover (or crimson clover or birdsfoot trefoil) is about fungal progression.[38] These particular legumes have a stronger affinity for mycorrhizal affiliation than do white and yellow clover species. The final stage of the cover crop plan above featured a mix of oats, red clover, and a modicum of tillage radish. The oats serve as a "nurse crop" in sheltering the smaller clover seedlings; tillage radish drills deep and takes up its share of the space between clover plants. Oats and radish alike will winterkill in growing zones touched by deep chill — leaving extant clover root systems hither and thither where tree holes will be dug. No worries there. Similarly, it's no big deal to fork out clover taproots in soil with good tilth where raspberries are about to go. Those clover plants more than an arm's reach away can remain, to continue to fix nitrogen and attract pollinators.

Another approach to fungal prep calls for sheet mulching at each tree location. This can really suit in rough terrain where breaking ground would only result in broken equipment. This permaculture technique takes a minimum of several months to be successful, and more like a year, if not two, if dealing with aggressive rhizome grasses. Spread compost or barn muck a generous arm's length around each tree location as a base layer. Adding a pound or so each of Azomite, kelp meal, rock phosphate, granite meal, and gypsum (whatever you got!) atop the compost contributes a bounty of minerals. Cover with corrugated cardboard or thick layers of newspaper to block all light to the sod below. Finish by piling up ramial wood chips over the cardboard a humongous hand's length deep. Rains will follow, and all that goodness will slowly meld into humus-rich soil. Mycorrhizal inoculum is not applied beneath sheet mulch, as roots are required for the symbiosis affiliation to engage; so wait on that step until tree-planting time.

Lastly, the nuanced improvisation that will secure your doctorate in soil prep: Stirring ramial chipped wood across a field expanse in the fall months jump-starts fungal connections. The researchers in Quebec were especially focused on this *thin layer transition* toward long-term fertility.[39] Freshly chipped twig wood from deciduous trees (post leaf fall) is spread an inch thick on average.[40]

Sheet mulching with cardboard, stable muck, and ramial chipped wood a year ahead of planting makes for easy digging when the time comes to set the tree.

These must not be plowed under, but rather lightly incorporated into the top 2 inches (5 cm) of cover crop residues and soil. No deeper, mind. The fundamental mechanism relies on massive entry of soil organisms into the chip-exposed cambium. The autumn period favors the proliferation of white rots. Basidiomycete species remain active at temperatures below freezing, whereas many bacteria go dormant during the cold season.[41] A ground spray rich in unsaturated fatty acids ensures this turn of events. Orchardists have overlooked this part of the ramial research too long. Plans for a high-density planting can proceed that very fall or come spring with the satisfaction of knowing the entire field has gone fungal.

RAMIAL PRUNINGS

Fruit trees are pruned to open up the canopy to airflow and sunlight. New wood in the form of shoots replaces older bearing branches, thereby rejuvenating fruiting potential within the allotted radius of the tree. Sometimes it's simply about keeping the height of an apple or pear tree within reasonable reach. Whatever the reason, whatever the season, the branch removed needs to be dealt with. Time yet again to choose between human notions of neatness or biological enrichment.

I was one who for many years gathered up prunings on a tarp in early spring and hauled the branches from the orchard to a burn pile. The result

Prunings from deciduous fruit trees certainly qualify as ramial wood. Cut small-diameter branches into short enough pieces to essentially lie flat on the ground.

was carbon in the sky and some mineralized ash left behind. No gain biologically, but the orchard understory sure looked tidy. Once I understood how ramial wood chips are a perfect fungal food, branches got chipped in place, essentially scattered under the dripline of the apple trees. Initially, this job got done using a home garden chipper with a narrow infeed, which required that crotchety tree branches be further cut down in order to fit into the hopper. Silly, especially knowing the insights that lie just ahead. My purchase of a 35hp tractor some ten years ago included a heavy-duty Wallenstein chipper that runs off the power takeoff (PTO) of the tractor. Branches up to several inches in diameter self-feed into a hopper mouth about 3

feet (1 meter) square, with the sidewalls sloped to bend side limbs onward for the most part.

Regardless, feeding branches into a mechanized wood chipper is hard work. You're in the midst of diesel fumes and noise aplenty. Larger branches are occasionally tugged violently from your hands. And oh, that incessant bending over to pick up watersprouts and smaller limbs. You're in a battle zone, with ear protection, given a singular focus. A guy loves this sort of thing! Or do we? Certainly the goal of ramial wood chips for the fungi rocks the boat . . . but then I got to thinking.

Definitive caveats determine what works best for soil biology. Prunings from deciduous fruit trees certainly qualify as the right sort of species primed for white rot action. A greater proportion of cambium nutrients will be reclaimed from watersprouts and smaller limbs (less than 7 centimeters in diameter, or not much more than 2½ inches across at the large end) relative to the heartwood within — and the majority of what gets pruned from well-managed orchards is surely so.[42] Branch pieces cut short enough to lie flat on the ground will be decomposed by soil organisms as opposed to aerial decay.

With all that in mind, I began cutting side limbs off the smaller pruned branches so each piece would come into contact with the ground along most of its length. Invigorated watersprouts can be dropped outright provided these lie relatively flat. Cutting a branch into shorter sections might seem tedious, but I find it flows into the rhythm of pruning, especially knowing I will never pick up such wood again to ram it through a chipper. I still throw bulkier branches into the aisleway to be chipped later. That said, a home orchardist can certainly choose to cut pieces 4–8 inches long to keep the fungal duff zone looking up to landscape specs.

Let's keep some ramial perspective from an earlier discussion on nutrient-rich cambium in mind (see "Soluble Lignins," page 72). Dealing with prunings in a nonpowered manner — without a chipper — simply means the nutrients in smaller wood won't necessarily be available as readily. The bark barrier on an extant branch slows down the entry of soil microorganisms. The larger the diameter of those branch pieces, the more time for drying out and seeing the soluble lignin advantage dissipate.[43] I do some of both in my orchard, thereby investing in fertility banking across time.

And let's not overlook that different decomposition environments have different casts of characters. Desirable saprotrophic fungi take care of organic matter on the ground, slightly buried

or not, typically within the span of one year. Conversely, a branch "hung up in the air" can take several years before collapsing down to soil level and a proper end. Decay fungi working in aerial environments can include parasitic species that might cause other problems in the orchard and the berry patch. Losing sight of decomposition nuance has led to advice to remove cankered wood and suspect prunings (such as spent raspberry canes), which, frankly, encourages far too much fuss.

Growers familiar with *Botryosphaeria obtusa* know "black rot of apple" in one of three forms: rotting fruit, frogeye leaf spot, and as limb cankers. Mummified fruitlets, fire blight infection sites, winter-injured wood, and dead branches left in the tree serve as the primary sources of black rot inoculum. Piling up prunings on the edge of the orchard without bonfire intentions is a bad call.[44] Branches with obvious cankers can indeed be removed and shipped to Idaho (just kidding!) but there are better biological options. Chipping, for one. The black rot fungus simply does not dance in a soil ecosystem; thus there can be no such thing as *Botryosphaeria*-infected wood chips in an active decomposition zone. The organism dynamics have changed. Worrisome branches can also be spread across low places and then covered with mulch hay or woodsy compost. Think hugelkultur on a pocket-sized scale. Organic matter with black rot potential or a raspberry virus or what have you can be kept to build soil . . . provided such branches or canes become part of this good earth sooner rather than later.

One bacterial woe deserves special mention in this regard. Fire blight often proves devastating to apple, pear, and quince because this disease can kill the tree itself. Bacteria of *Erwinia amylovora* find openings into the vascular system of the tree through exposed blossoms and succulent shoot

Black rot cankers originate from deadwood in the orchard environs. Photo by Liz Griffith.

Commercial orchardists often pile pruned branches in the aisleway to be flail-chopped. The same principles apply and this is far quicker work, if not quite as ideal as directing chips into the fungal duff zone beneath the trees. One way or another, prunings from deciduous fruit trees can find the way to ramial glory.

PULSING THE SPRING ROOT FLUSH

Apple and pear trees actively extend their feeder root systems twice during the growing season.[46] The *spring root flush* kicks into action immediately after blossom time, just as fruit begins to set. Nutrient uptake at this point in late spring gets split between seed formation within developing fruitlets and investment in blossom meristems that will become next year's crop. The *fall root flush* begins by the end of August, once terminal buds on shoots actively stop growing, thereby initiating hardening off of cambium tissues in preparation for the winter ahead.[47] Nutrients stored in buds and inner bark throughout the fall months will drive green-up come the following spring.

Feeder roots extend an inch or two at most off the permanent root system of the tree. The majority of these lateral extensions are temporary, given a particular job to do, and then shed.[48] Action down below occurs when shoots are not actively growing up above, which makes sense, given competition for growth energy between root outreach and shoot extension. The bulk of feeder root action takes place in the humus layer beneath the tree and extending a couple feet beyond the dripline of the tree — what I have deemed as the *fungal duff zone* for holistic orchard management purposes.

This ebb and flow of feeder root growth is met with a similar response on the part of mycorrhizal fungi. Hyphae broaden the reach for nutrients just

tips. Leaves on green shoots turn brown or black and bend over into a characteristic shape not unlike the top of a shepherd's crook. Shoot blight infections extending down the stem enter older wood to overwinter as cankers.[45] Sticky bacterial ooze seen along the edges of such cankers on early-spring days brings a next round of fire blight deployment to infect anew. Activated cankers are a risk for a short time even off the tree, especially if pruning once growth has begun. Common sense suggests these are the branches to remove from the orchard environs and immediately bury or burn (or ship to Idaho!). Pruning in the winter months doesn't necessarily demand such promptness, just as long as cankered branches get chipped or removed before the weather warms.

as the tree is capable of delivering more sugars in trade. Early shoot growth has brought more leaves into play to photosynthesize, which in turn provides spring feeder roots with more carbon to trade with fungal partners.

Timing the first mowing of the orchard understory to this rising tide creates favorable dynamics for the uptake of nutrients for growing fruit. Trees planted in a meadow-type ecosystem face competition from surrounding grasses and herbs.[49] Scything down this sward in the immediate week or two when fruit has begun to set, round the allotted radius of each tree, places a carpet of mulch with an ideal carbon-to-nitrogen ratio for fungi to thrive.

This fungal sweet spot occurs when carbon levels of applied organic matter are thirty to forty times higher than nitrogen. A quick lesson in cutting green hay explains why. Farmers traditionally make the first cut of hay when protein levels are optimal for livestock. This coincides with when grasses start to form seedheads. The growth cycle of oats shows this nicely: "Lush and green" early on indicates higher levels of nitrogen; the developing grain brings out yellowing highlights in oat leaves; fully mature grain heads wave in the breeze atop carbon-rich oat straw. The most desirable stage for first-cut hay comes when the seedheads of grasses are still immature.[50] That stage provides the right carbon-to-nitrogen ratio to prod abundant fungal support, beginning, in the case of the orchard, when most of the apple blossoms have fallen from the trees.

It's important that dripline mulch be unchopped, laid down as full-length stalks. Scything achieves this well.[51] Commercial orchardists with a considerable number of trees can use a sickle bar mower to drop the sward alongside each row soon after petal fall. Growth suppression of surrounding plants gives feeder roots of fruit trees

a longer chance to interact with soil food web allies of all persuasions.

Down below, root shock to these plant competitors provides *room in the humus*, so to speak, for tree feeder roots to access the greater share of nutrients in more complex bioavailable forms. Infrequent mowing has an entirely different effect on grasses than does regular mowing.[52] This first knockdown around the trees in late spring actually induces net pullback in neighboring roots until those mowed plants recover. Mycorrhizal density increases as tree feeder roots take advantage of nutrient release in discarded rhizospheres. This in turn ensures healthy plant metabolism for fruit trees at a critical juncture.

FATTY ACIDS IN ORCHARD SPRAYS

Many challenges await fruit trees. Orchardists build on the biological advantage trees reap from healthy soil through the application of holistic sprays keyed to site dynamics. The "primary infection window" in spring is the most paramount of all — scab spores, rust spores, and first-stage rot spores land on leaf and blossom surfaces from green tip into the fruit-set period.[53] Depending on the fruits you grow, the place you be, and the weather (oh, indeed, the weather!), disease concerns can continue throughout the summer months.

A guy writes books to explain all this in full detail. Those of you familiar with *The Apple Grower* and *The Holistic Orchard* know the drill. Our purpose here is to zone in further on the usefulness of fatty acids in an orchard context.

Mycorrhizal and saprotrophic fungi in the soil respond to lipid-based food resources, which can prove especially useful when the ground begins to warm in early spring.

Fatty acids provide sustenance to arboreal organisms, including the single-celled fungi known as yeasts. Competitive colonization is integral to keeping disease from getting a foothold on plant surfaces.

Perennial canker infections can be checked by means of "fat nutrition," providing a leg up to competing organisms on bark tissue.

The healthy sheen left on bark and leaf alike helps protect plants from abiotic stresses, including extreme cold and parching winds.

The ability to disrupt pathogens from overwintering in bud and bark crevices holds particular promise for growers faced with bacterial spot, peach leaf curl, and ever-worrisome fire blight. Such a "fatty acid knockdown" involves a flood of fats to soften the lipid coating protecting the disease organism, followed by a flood of competing microbes within twenty-four hours to change the dominant paradigm.[54]

The primary sources of fatty acids in holistic orchard sprays are fish hydrolysate and pure neem oil. Liquid fish can be used up until midsummer, after which you are well advised to "lay off fish" while trees begin the hardening-off process for the winter months ahead.[55] Pure neem oil gets applied at different concentrations depending on the targeted purpose.

Early-season neem goes on at a 1 percent concentration when used as a fungal catalyst directed at the ground and major branch structure of the tree. Little leaf tissue shows when this first spray gets made, thus the risk of phytotoxicity is low. The azadirachtin component of neem allows holistic growers to skip applying "dormant oil" at this time. Petroleum-based horticultural oils smother the eggs and nymph stages of aphids and mites that overwinter in buds and bark crevices. Neem works

instead by putting a halt to molting (instar development) of these foliar pests.[56] One additional bonus applies to codling moth. This major fruit pest overwinters as larvae (resting between seasons in what's known as "diapause") behind coarser tree bark. Skin saturation follows bark saturation, thus lessening moth pressures.

This same concentration of neem (and even up to 2 percent) applied to the trunk zone in the summer months deters various borers. Indeed, soak the ground until the neem spray puddles up . . . this ensures complete bark absorption at the soil line

Spraying to augment arboreal biology and plant nutrition is a whole different ball game from spraying toxins to address pest symptoms caused by biological and nutritional deficiencies.

and thus a premature end to any grubs within. I never worry about soil life in the vicinity as *neem oil = fatty acids = healthy diet* from the perspective of mycorrhizal fungi.

All other foliar applications of neem oil in the growing season are made at a 0.5 percent concentration.[57] Remember to avoid spraying when temps climb into the 80s (above 27°C), as even this lighter rate can prove phytotoxic when overapplied under a hot sun.

Post-harvest neem gets put on at a recommended 1 percent concentration to bark and ground alike during the leaf abscission period. One of many objectives behind the fall holistic spray includes mycorrhizal uptake.

BRIDGING THE AM-EM DIVIDE

The synergy that unfolds as a result of outrageous diversity in the orchard delights me to no end. All sorts of flowering plants bring all sorts of beneficial insects to protect the fruit above. Tree root systems below "share the humus" with taprooted herbs mining further down for subsoil minerals. Feeder roots able to access partially built nutrition in a thriving food web contribute greatly to a tree's innate resistance to disease. Yet it's only through mycelial diversity that fruit trees receive the full blessing.

Multiple species of mycorrhizal fungi network beyond our imaginings in the underground orchard economy. We learned about this underlying intelligence earlier when discussing *anastomosis*, whereby hyphae of the same species merge networks. Such "fungal grafting" allows otherwise unconnected plants to share nutrients, water, and defense signaling. Similarly, a *passage plant* (sharing root affiliation with two species of arbuscular mycorrhizae) becomes a conduit for enhanced trade, taking

in an excess nutrient only to send such back out on a different hyphal pipeline.[58] But here's the thing: Most fruiting trees and those deep taprooted plant allies lack ectomycorrhizal affiliation. Arbuscular mycorrhizae only reach so far beyond the nutrient depletion zone of the root, a marvel in itself but not the whole shebang.

Certain tree species form mycorrhizal associations with both AM and EM fungi. And it's through having these noncommercial species in the environs of "arbuscular fruit trees" that long-distance hyphal trade comes into play. Ectomycorrhizal fungi alone have the wherewithal to bring nutrients and water from afar. Those *mycorrhizal bridge trees*, which rely upon AM and EM simultaneously, are a conduit for nutrient exchange yet again. This brings outreach capabilities to fruit trees heretofore unconsidered in the herbicide monoculture of conventional orcharding. Community-based orchards on a smaller scale likely benefit from forest-edge proximity if the right sorts of trees are nearby. Permaculture polycultures occasionally achieve this mycorrhizal bridging by happenstance. The time has come to straddle the AM-EM divide.

The root systems of fast-growing trees with relatively pliable wood make barter possible between AM and EM fungi. Alder, aspen, cottonwood, poplar, and willow are chief among the bridge trees that take orchard health to an entirely new level. Most of these are members of the family Salicaceae, found throughout the temperate parts of the world, with the majority of species occurring in the north. Both willows and poplars have a strong affinity for water, and can commonly be found near ponds and along watercourses. Alder comprises a genus of some thirty species belonging to the birch family, Betulaceae. These trees and shrubs, cloaked in catkins each spring, improve the fertility of the soil where alder grows by means of

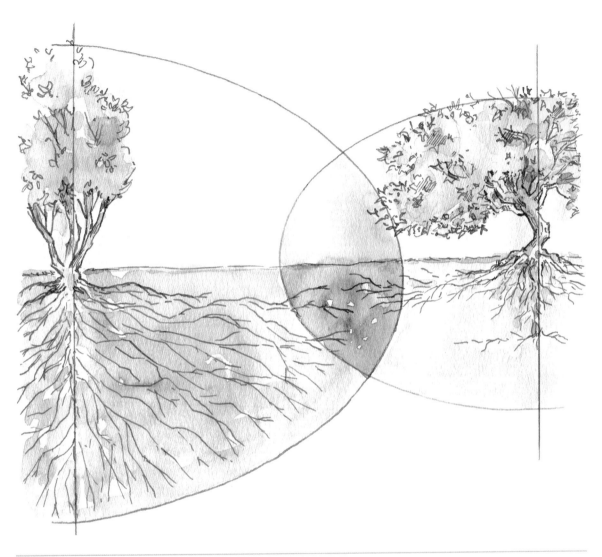

The long reach of tree root systems enhances mycorrhizal network connections. Alder and apple alike extend further than you might expect. Adapted from Robert Kourik's *Understanding Roots*.

a symbiotic relationship with nitrogen-fixing bacteria.[59] A number of species of eucalyptus can be included here as well, noting at the same time that eucalyptus trees release allelopathic compounds that inhibit other plants.[60]

Both willows and poplars have very vigorous root systems, stretching up to 130 feet (40 m) out from the stem. Red alder and black alder are serious trees, reaching a height of 70 feet or more (20-plus m) at maturity. The swamp alder and gray alder are smaller, often reaching not much more than 15–20 feet (up to 6 m) in height before giving up the ghost to a younger stem. That's more workable with respect to orchard proximity. Such shrub alders can be coppiced down low every few years as these root systems will almost always sucker anew.

The ability of alder to affiliate with both AM and EM fungi makes it a good bridge tree to have near fruit trees.

of *hardwoods* and *softwoods* earlier in distinguishing the right sources for ramial wood chips — based on an entirely deciduous differentiation — this grouping of mycorrhizal superstars are better described as *soft hardwoods* by cabinetmakers. In fact, poplar species rank among the very softest woods in the world.[61] All but eucalyptus are indeed deciduous. Which got me to thinking about other "soft hardwoods," where perhaps scientists have yet to affirm the arbuscular side of the divide. The lesser maples are a case in point.[62] Fungi evolve just as trees evolve. One way or another, this is diversity seen in an entirely new light.

FUNGAL COMPOST POINTERS

Compost is not "any ol' heap of organic matter" when you have biological perspective.

Bacterial compost results from a thermal process featuring a higher nitrogen charge and several turnings of the pile to aerate the microbe scene. Both of these actions rouse bacteria into action. The thermophilic phase in particular gets the pile cooking, thereby cleaning house of potential pathogens and weed seeds. The whole process generally takes six to ten weeks from start to finish. Commercial compost operations make this sort of "garden compost" for certified organic growers (adhering to federal restrictions) and home gardeners purchasing compost by the bag.

The carbon-to-nitrogen ratio of the materials to be composted expresses how nitrogen stands in relation to carbon content. Keeping the C:N ratio closer to 25:1 promotes rapid decomposition. A simple recipe for bacterial compost involves alternating layers of green (nitrogen-rich) and brown (carbon-rich) materials in equal proportion.[63] Adding a charge of fresh manure from livestock or poultry to these piles ensures the proper nitrogen

Talk about a regenerative source of nearby ramial wood! Mycorrhizal bridge trees can be integrated into the orchard environs from relatively afar, given the reach of roots and the ectomycorrhizal half of the equation, perhaps as much as several hundred feet. The interconnecting mycelia shared among arbuscular understory plants between bridge trees and fruit trees are vital to making a more global connection possible. There's a sensibility to this, given that smaller renditions of willows and alders like wet feet and fruit trees do not.

The woods of bridge tree species are lighter and more porous than other woods. Whereas we spoke

The Uncertainty Principle

The possibilities for mycorrhizal bridging seemed about right to me . . . until I chanced upon a paper written some thirty-odd years ago by Jack and Lindsay Harley titled *A Checklist of Mycorrhiza in the British Flora*.[64]

Listed are references to mycorrhiza affiliations to native plants of the British Isles observed throughout the previous century. The authors clearly state that while some of the research can be trusted, "most of the entries are misleading, untrue by omission at worst, or misleading at best." This actually just made me *curiouser and curiouser*. Moreover, the fungal world does exhibit fluidity depending on the season, environmental conditions, soil type, and phase of root development. There were some surprising additions to consider as possible bridge trees based on shared association with arbuscular and ectomycorrhizal fungi.

Hawthorne (*Crataegus monogyna*)
European crab apple (*Malus sylvestris*)
Wild cherry (*Prunus avium*)
Wild pear (*Pyrus pycaster*)
Red-berried elder (*Sambucus racemose*)
Rowan (*Sorbus aucuparia*)

Are you digging the implications of all this? Might we be able to establish mycorrhizal symbiosis across the AM-EM divide in our orchards through traditional crossbreeding? Or is this more about localized ectomycorrhizae with a British accent?

Perhaps the real lesson here is that certitude can be *rethunk* at any point in time.

fix. Turning the pile weekly—as many as five times—allows bacteria to stay the course in all that biomass. Bacterial compost is just the ticket to help annual vegetables and flowers thrive.

Every compost pile of mine brings in a fungal element from the get-go. This "partial static approach" to thermal composting involves but one round of turning, and several months down the road at that. I steer the same course to start, layering nitrogen sources with carbon sources, and include dustings of kelp meal and Azomite for minerals. The inner core of a diligently made static pile heats up, while the outer edges take on a fungal imperative.[65] Leaf litter from the forest floor brings a diverse fungal prospectus, in addition to making nice wadding to fill in against the stacked logs that enclose each pile. Fatty acid sprays (made whenever passing by on orchard runs) add a lipid boost for the decomposers working the periphery.

I typically start a new pile every spring, another by midsummer, and at least one fall pile that will carry through the winter months. The fixings come as we do garden work, mow grass, put up produce, rake leaves, press cider (thus creating pomace), and muck out animal bedding. Four to six months later comes reckoning time. This somewhat immature

Stacked-log composting works well for building up green and brown layers over the course of a few months. These "bins" can readily be disassembled when the time comes to move compost onward.

compost can be moved in one of two directions after taking away the stacked logs to provide access. Compost for garden use most often gets piled close to where a cover crop rotation is on deck or where garlic will be going in late fall. Bucket loading by tractor provides a rough-and-tumble mixing of the works. Give these field piles a few weeks to settle and cure. The resulting organism-rich compost is ready at that point to spread across field stubble.

Compost destined for orchard use faces another leg on its journey to application. Orchard piles consist of one part "rough-and-tumble compost" mixed throughout with one part ramial chipped wood, thereby upping the carbon-to-nitrogen ratio to more like 40:1. Fungal sweet spot territory. Black gold. Texas tea.[66] This woodsier pile will require another four to six months to come into its own . . . as the spot-on *fungal compost* craved by fruit trees and berries alike.

A few important distinctions should be made about what's taking place at this juncture on the biological time line. Fungi thrive on nondisturbance. Orchard compost will not be turned going forward in order to allow white mycelia to develop and spread throughout the bulked-up pile. That partial static variation on thermal composting makes possible a wider diversity of beneficial fungi than might otherwise occur. The ramial chipped wood helps create air passages that provide even more oxygen for those aerobic fungi to flourish. Soluble lignins in hardwood chips will begin the journey toward humification, ultimately forming macroaggregates. Conversely, using somewhat-aged softwood chips to create an orchard pile puts white rots at the helm in reclaiming more carbon than not.[67] All organic matter has value in some form or another.

The passage of time is what's key. Protozoa and nematodes reactivate during this maturation period. Microarthropods increase in number. In fact, once the pile has cooled down, diversity actually will continue to improve for the next six months.[68] Orchard piles made in spring have additional potential if positioned along the forest edge. Tree roots reaching into the pile will leave behind mycorrhizal hyphae and spores by the time of late-fall application.[69]

Why fall? Compost applied toward the end of leaf abscission furthers two orchard aims. The fall root flush continues to be at full bore — along with mycorrhizal outreach — and thus it is perfect timing for bioavailable nutrients from fungal compost to hit the ground running. Enhancing leaf decomposition ties in here as well. Compost anchoring down the fungal duff zone as the trees head into winter introduces yet more diversity. And should winter come early, take heart. Spreading fungal compost in the orchard and around berry plantings in early spring has virtue, too.

Leaf Decomposition

Final credit goes to saprotrophic fungi and earthworms for the disappearance of fallen leaves beneath fruit trees.[70] Extracellular enzymes secreted by both release nutrients from the organic matter on the ground surface. These nutrients are available for uptake by saprotrophic and mycorrhizal fungi alike, along with tree feeder roots infusing the fungal duff zone. So why would anybody want to remove leaves from the orchard?

Scab, in a word. *Venturia inaequalis* on apple; *Venturia pirina* on pear. These disease fungi cause black spot on apple and pear, overwintering primarily on infected leaves on the ground. (And as twig lesions as well in the case of pear scab.) Some growers believe that it's best to remove leaves, thereby considerably reducing disease potential long before ascospores are released the following spring to cause infection anew.

Which can be quite a job in a community-scale orchard. Digging deeper into the facts reveals a more biological approach. Decomposition of leaves can be enhanced by means of a fall holistic spray to make the leaves that much more palatable.[71] Those fatty acids along with nitrogen in the liquid fish are foods for the masses. Mowing in fall to reduce vole cover results in many fallen leaves being chopped into more digestible bits. Intact leaves infected by scab over the past growing season will be colonized completely by the mycelium of the pathogen within a month after falling from the tree.[72] Spore sacs will have formed by the time spring comes around. Leaves blown about by a mower at this time — or simply raked within the fungal duff zone beneath each tree — get tossed and turned, many upside down, thereby pointing quite a few ascospores in the wrong direction for effective release.

Taking away organic matter from a living soil misses the mark. Nature builds soils by reinvesting those fallen leaves. Litterfall is a dominant pathway for nutrient return to the soil, especially for nitrogen and phosphorus. Nor should we forget the litter decomposers, those organisms whose niche is solely to decay leaves. These particular fungi and bacteria employ different strategies to gain access to foliage substrates. Take note especially of *Microsphaeropsis* fungi, which seek out scab spores to eat like caviar.[73] Losing what's provided free of charge in the name of excessive hygiene is nothing more than ass backward. We can deal with apple scab and pear scab far better by boosting tree immune function.

Forest

The ridge beyond our farmhouse is classified as a "ring dike" that was formed long ago by an upwelling of molten magma in a circular fracture of the Earth's crust. Many of the unique plant communities in what's known as Cape Horn State Forest are protected by the steepness of the terrain and the black spruce bogs below. Nancy had the idea to hike up to the ridge to celebrate a recent Thanksgiving in the woods. We headed out the door, golden retriever in the lead, and followed the winding brook to the basin of the horn. The hike up the eastern slope brought us to a mesic upland forest consisting primarily of sugar maple, white ash, beech, and basswood. The species diversity is greatest in the herbaceous layer, with some of the most abundant species including northern maidenhair fern, blue cohosh, wood nettle, zigzag goldenrod, Clayton's sweet cicely, rattlesnake fern, spikenard, and plantain-leaved sedge. Rare species such as *Dryopteris goldiana* (Goldie's fern) and *Sanicula trifoliata* (beaked sanicle) can be found as well. The mycorrhizal network weaving all these plants into a harmonious whole was the blessing underfoot to an absolutely stellar day. We wedged ourselves at the base of a beech tree to share a backpack feast and give thanks for this life.

The final pitch to the ridge was slower going but irresistible given the panoramic view of our northern New Hampshire home. The north peak lies just beyond the state forest boundary . . . and that's where our day changed completely.

Loggers had been at work just recently, as we knew from the roar over the ridge that fall. The privately held lands on the downslope toward town weren't entirely bereft of trees but they were destroyed at soil level by oversized equipment. Monster ruts nearly 2 feet (60 cm) deep had been made to remove 9-inch-diameter (23 cm) stems for firewood and biomass chips. The "whole tree highway" leading down the ridge was seemingly designed to remove whatever fungal loam remained. That this might be considered good forestry in the context of what's happening elsewhere in the world — islands of small trees were still standing, after all — does not begin to justify doing undue harm to any ecosystem.

Chaos rules the future of most of the earth's forests today. Be it the pressure of paying off bank loans for massive equipment that results in lowest common denominator forestry or insatiable greed on the part of investors financing the removal of rain forests for palm plantations in Malaysia. Clear-cutting vast tracts is considered to be economically efficient, which, followed by burning slash and herbicide application, leaves only a lifeless paradox behind. The current rate of deforestation plays hand in hand with conventional agriculture in doing exactly what's required to shut our species down.

We need to heed the perspective of the fungi to put things right. Trees stand in a "fungal forest" where as much as a third of the biomass in the topsoil may be mycelium. The concept of *mycoforestry* involves far more than merely inoculating replant seedlings with ectomycorrhizal fungi. A mixed-age forest has many more components than timber alone. Granted, *anything* we do in the forest will

All species contribute to the health of the forest, from the trees and understory plants to the fungi and other allies in the soil.

have a detrimental impact to some living system. Our task is to minimize that impact by choosing diversity over short-term gain.

A sustainable forest economy only becomes possible if we start to reckon every cost, be it unraveling mycorrhizal networks or the sheer stupidity of undoing the biotic pump that delivers rain to inland regions.[74] Solving the overreach of industrial forestry will take concerted effort. The audacity of replacing biodiverse forests with tree plantations will not deliver us from evil. The question of who owns the land will determine how far we can go.

Community forests — owned by the people who live nearby and have reason to care about healthy trees and watersheds and exceptional species — become refuges of sanity where good forestry can be not only demonstrated, but sustained for generations to come. Woodland acreage removed from the commodity pool can be managed for recreation, wood quality, mushroom foraging, and just because trees matter. Unique groves can be preserved and old-growth trees celebrated. Kids and grandparents alike can go on a "mushroom tour" and learn about fungal wonders. Moss grows back when tree stands have a mix of ages and species. Harvest management plans can be set out in increments of centuries. All can be directed by what is best for the forest ecosystem as a whole.

Cortes Island lies between Vancouver Island and the mainland of British Columbia, Canada. There the Wildstands Alliance has taken on an industrial logging machine bent on mowing down island forests at an accelerating rate for the purpose of raw log export. It's the sort of corporate mind-set that operates the world over for the profit of the global asset shareholder. On Cortes, it faces a community of a thousand residents that has fought the cashing in of forest integrity for the last two decades with a

high level of success . . . to the point that corporate heads have dubbed the islanders as being "socially inoperable." A badge of honor, that. Outright resistance to the short-term goals of wealthy people who would destroy the ability of the biological world to renew itself is needed everywhere.

Fungi Perfecti has purchased 160 acres on Cortes Island to conduct its mycoforestry research project. The land itself had been managed as an even-age stand; the red cedar was cut, leaving behind pockets of Douglas fir. Paul Stamets and crew started by chipping the slash and thus utilizing important organic matter in reestablishing the nutrient base of the soil. (This as opposed to the usual burning of slash piles made to knock back disease potential.) The blades of the chipper were coated with mycelium in order to jump-start fungal resiliency. Sitka spruce and grand fir were included in the replanting scheme to broaden species diversity. Root systems were dipped in liquid mycorrhizal inoculum — admittedly more labor-intensive than throwing a "myco tea bag" in each planting hole — but then again, bringing biological emphasis to forest restoration calls for going the extra mile. The standing firs serve as adoptive mother trees. The growth of transplanted seedlings has proven phenomenal as a result of reestablishing mycelial support so quickly. Such simple things are first steps in ensuring the return of a varied forest canopy. Cortes Island has yet another piece of its forest legacy preserved for future generations.

HEALTHY SUCCESSION

Managing a forest properly for timber harvesting results in relatively little harm done to the common mycorrhizal network. Meet that definition and you are doing things right, whether as a forester, logger, or woodlot owner.

Property lines do not reflect forest ecosystems as a whole. Every stand of trees is part of a larger landscape. Thinking about areas much larger than foresters normally consider helps in evaluating habitats for indigenous species. Maintaining the "integrity of the forest" often goes well beyond the chunk one human claims at any moment in time. David Perry of Oregon State University has been speaking to this big-picture view for many years. His book *Forest Ecosystems* reveals the complex patterns and processes observed in tree-dominated landscapes. Needless to say, applying a fungal framework to forestry practice would greatly change the ways things are currently done.

Conventional succession reflects the integrity of a stand as it seeks to recover from disturbance. Let's phrase this in terms of human-induced forest management . . . noting that we really can't do anything about hurricanes and other natural phenomena. Foresters name four stages in describing how the forest ecosystem comes back from disturbance: stand initiation, stem exclusion, understory reinitiation, and steady state. Each stage has specific characteristics that are typical regardless of the type of forest that has been logged.

The plant response to sunlight streaming down on the opened forest floor is vigorous, to say the least. The trees that will establish for the next harvest interval do so immediately—along with many, many other stems. The duration of stem initiation is relatively short, on the order of fifteen to twenty-five years. Seedlings, stump sprouts, and transplanted nursery stock compete to stay in the light and become the dominant players. Management in the stem exclusion phase consists of thinning less desirable species and stems with forked tops. Some twenty to forty years later a tentative balance begins to emerge. A percentage of trees will be mature enough at that point to set

seed; dappled sunlight again reaches the ground; understory species take root. Early successional dominants such as poplar and paper birch will be of sufficient size to harvest should a selective cut be made. High-quality stems could be given more time . . . but this is where economic temptation often steps in and the developing forest gets sheared anew to pay the finance mobsters. The clear-cut becomes the steady state, so to speak, when waiting a hundred years for a partial return is viewed as untenable.

Let's consider what the fungal network has been doing during these first decades of succession. Much was lost at the time of the initial cut. Damage from heavy equipment subdivides the once intact mycelium into fragments across the woodland ecosystem. Hardwood stump sprouts attempt to carry things forward by means of relatively few leaves (and incredibly oversized ones at that) to much-reduced root systems. Seedlings lack the mycorrhizal nurture of mother trees, but odds are stray ecto spores will slowly bring what will become the fastest growers back into the fold. The loss of diversity in the herbaceous understory lessens critical nutrient sharing between species. AM fungi have a nearly impossible time recovering if the herbicide card is played to start things off. Resiliency in an even-age stand is put on hold indefinitely.

All shifts when working with healthy succession in a mixed-age stand. The mycelial network can retain connectivity across the ecosystem when we choose to be low impact and take the long view. Select legacy trees are left to live out their natural existence, thereby preserving ties across generations. Young trees are provided with carbon sugars and other nutrients while struggling to stand on their own two feet. Those stump sprouts expected to maintain fragile fungal ties in a clear-cut get support in turn, thereby strengthening the network as

Talk about how a little dab will do ya! The inoculated nursery tree on the left is leagues ahead of its companion in terms of both root mass and needle vigor above. Courtesy of Dr. Mike Amaranthus, Mycorrhizal Applications.

a whole, while new trees fill back holes made in a selectively cut canopy. Big deadwood becomes a sanctuary for invertebrates and more microbial diversity. Don't lose sight of the fact that ecto-mycorrhizal fungi draw nutrients directly from organic matter, including successional species left to decay on the floor of the forest. Fungal support during periods of intense water and nutrient stress keeps even the towering redwoods in good stead during hard times.

Aldo Leopold had it right when he said *keep all the pieces*. A clear-cut in no way mimics a natural disturbance. Whole-tree harvesting opens up the canopy but takes away the legacy that nurtures the next round of trees. Loss of organic matter through removal of all biomass (other than decaying roots) and burning of slash undermines tomorrow. More than that, bacteria and fungi teamed with invertebrates make up 90 percent of forest species. Diversity favors the proportion of predatory insects to tree-feeding insects. The unseen are what provide true resilience in any ecosystem. Prudent behavior demands that we do not lose any species in managing forests. Mixed-age stands accomplish what even-age stands never get to ponder. Once you've cut everything in sight, the option to have big trees is gone for one hundred to two hundred years, depending on where you live. Leaving some big trees keeps all possibilities on the table, and certainly cheers the human heart.

Forest successions face additional situational challenges where mycorrhizal affinity always proves advantageous.

Clear-cuts in the West often require replanting, which in turn requires nursery stock to make appropriate reintroduction of desired tree species. Fungal inoculation wasn't always a given. Mike Amaranthus launched his career with the US Forest Service as a soil scientist at the Rogue River–Siskiyou National Forest in Oregon in the late 1970s. He noticed that the bushy Douglas fir and other conifer seedlings that came from the nurseries often didn't do well when planted out on harvest sites. Attending a lecture at Oregon State University on soil microorganisms brought it all into focus . . . and soon the man who would launch Mycorrhizal Applications some twenty years later began experimenting with inoculating tree seedlings. The results were positively enticing, with the more vigorous top and bulkier roots always in the mycorrhizal camp. More to the point, nursery seedlings, grown in sterilized soil and fed with

Garlic Mustard

Here's an underground perspective on invasive plants, in particular garlic mustard, currently proliferating in eastern woodlands. This plant's crime, if you will, lies in undoing mycorrhizal networks for deciduous trees. *Gasp!* says the gathered crowd.

Garlic mustard, *Alliaria petiolate*, appears as a rosette of green leaves close to the ground in its first year on the forest floor and then develops into mature flowering plants the next spring. A niche claimed becomes a niche protected. Garlic mustard produces antifungal compounds, which hampers the establishment of new associations between native trees and mycorrhizae. Sugar maples, red maples, and white ash exhibit less colonization of roots by arbuscular mycorrhizae. The effectiveness of existing ectomycorrhizal connections are diminished 4 inches (10 cm) at a time, radiating out from each and every new mustard taproot. Trees so impacted produce less biomass as a result of losing fungal support. Prospects for succession falter as next-generation saplings no longer have ties to parent trees.

Interestingly, garlic mustard does not wreak havoc with trees and other woody shrubs back on its home turf in Europe. Researchers believe this disruption of plant-microbe interactions is novel to North America. Species here may eventually evolve mechanisms around this. Still, nothing beats being able to eat your problems. Garlic mustard is a choice edible plant in its own right. A little olive oil, a little salt . . . and the mycorrhizal network gets back on track.

Garlic mustard gets an early start on the forest floor in spring, thereby securing its nonmycorrhizal niche at great cost to certain North American tree species.

fertilizer, were unable to handle the stresses of life once transplanted into the great outdoors, whereas fungal-affiliated trees could. Numerous studies would go on to show that *Rhizopogon* species in particular are aggressive colonizers in nonirrigated and harsh field conditions.

Trees certainly knew this from the get-go and accordingly have developed a brilliant strategy for propagation that works in conjunction with ectomycorrhizal fungi. What's known as a *mast year* takes place every two to seven years when all conditions are right. Meristem development the previous season is fulfilled; pollination is good; the sun shines when kernel cells divide followed by well-paced rains as kernel cells swell with lipids. These are the growing seasons when the seed crop proves prolific — be it acorns or cones, samaras or pods — leaving the natural world overwhelmed in sheer abundance. Animals take full advantage of the bounty but can't begin to consume it all. Some seedlings invariably succeed under this cover of plenty. More than that, such seasons cannot be counted upon year after year, which keeps animal populations on the move, opening up additional opportunities on the ecosystem edge. The fungal network nurtures these new kin until such time as the canopy above releases a niche. This evolutionary adaptation of mass fruiting at multiple-year intervals in random fashion would not be nearly as successful without the mycelial safety net.

Mycorrhizae can help trees of different species cope with climate change as well. Bark beetles, blight, and rust epidemics in pine and spruce are finding new niches as summers prove warmer and drier. Vast tracts of Douglas fir being hit hard by western spruce budworm lose the majority of their needles, leading to mass die-offs. This in turn opens up space for new species such as ponderosa pine, which are migrating north in response to climate change. Guess what? Suzanne Simard and her team have been able to show that mycorrhizae facilitate this transition by directly shuttling food from dying Douglas fir to young ponderosa pine. Both trees host hundreds of mycorrhizal species, sharing more than several fungi in common. Additionally, stress signals from the fir stimulate strong synthesis of defensive enzymes in the pine, should the budworm choose to turn its attention "pineward" instead. The species in decline offers its final reserves for the good of the new ecosystem. Mycorrhizal fungi carry things forward yet again.

UNDERSTORY CONNECTIONS

Far more is afoot in the forest than the trees overhead. Ecosystem resiliency ties to that "wood wide web," which only finds its full stride with understory diversity.

Case in point are the Ericaceae such as salal, rhododendron, and huckleberry, which are often part of a widespread dominant understory. These plants colonize woody substrates where ericoid mycorrhizal fungi have the ability to access organic nitrogen. This not only benefits the shrubbier hosts, but contributes to nitrogen cycling throughout the forest ecosystem.

Forests everywhere get disturbed to one degree or another, whether by humans or natural events. Recovery depends just as much on the comeback of understory vegetation as it does a next generation of trees. Work done at the Hubbard Brook Experimental Forest here in New Hampshire back in the 1960s made this crystal clear.[75] Herbicide treatments of early successional vegetation led to nutrients being lost on the order of fifteenfold — no surprise there, given that plants and fungi united as one are what hold soils together. Clear-cutting northern hardwood forests typically causes enhanced export of mineral nutrients such as calcium in drainage waters. Take out the small plants that make up the understory and organic matters just become far worse.

The flip side of this is where beauty comes into play. United Plant Savers maintains a goldenseal botanical sanctuary in southeastern Ohio. Life is just as much about the understory on these 370 acres as it is the noble trees above. Extensive botanical assays have been performed to determine the resources present on the land. Over 500 species of plants, over 120 species of trees, and over 200 species of fungi have been identified. Native medicinal plants are thriving in abundance on this land, including large communities of goldenseal, American ginseng, black and blue cohosh, and grand old medicinal tree species such as white oak and slippery elm. This UpS sanctuary has become a shining symbol for protecting diversity when people take a long-term view to forest stewardship.

Fungal diversity goes beyond tree affiliations, of course. The ectomycorrhizal relationship is to be expected, though it's not as predominant as might be assumed when looked at from the complete plant perspective. All the major forest biomes differ little in the percentage of plants that form relationships with arbuscular mycorrhizal fungi.[76] Temperate forests tally in at 56 percent, noting that species diversity here is often divided somewhat equally between trees above and a shade-tolerant understory. This becomes all the more true in tropical forests, where 70 percent of plants affiliate with AM fungi. The highly weathered soils of equatorial zones often provide little available phosphorus for plant growth, and given the rich species mix, this is entirely predictable (see "Fungal Adaptability," page 16). It's in the world's largest terrestrial biome — the boreal forest — where AM plant species affiliations are

Rhizopogon mycelium on conifer roots. Courtesy of Dr. Mike Amaranthus, Mycorrhizal Applications.

surprisingly high at 64 percent, on average. Those other plants in a coniferous ecosystem make the mycorrhizal network that much stronger.

Gourmet value in the fungal understory holds special sway for us humans. Many of the diverse mushroom species seen on the forest floor every summer and fall are the reproductive structures of ectomycorrhizal fungi. The notion of "mushroom tourism" as a way to take some portion of the forest out of the wholesale lumber stream ties into the desire by many to get back to Nature. Reasonable protection for mixed-age stands of trees requires an alternative economy to be successful, one that the fungi are more than willing to oblige.

Low-Impact Forestry

Behaving in the woods comes down to lessening harvesting impact on the forest floor and leaving a legacy of bigger trees in place for future generations.

Equipment dynamics determines much of this. A skidder or a feller buncher uses as much as 25 to 30 percent of the land area for access trails and yarding. That's one serious hunk of disturbance.[77] Moving trees to machine is less damaging to soil than moving machine to trees. A radio-controlled winch alone can cut back access requirements by a quarter to half of the surface area used for a timber harvest. Tracked vehicles exert less pressure on soils and can run over brush deliberately placed on trails to further minimize impact. Conversely, the wheeled feller buncher is the least desirable choice for respecting fragile soil ecosystems.

Efficiency of yarding is increased when logs are gathered into groups before yarding. The use of a grapple winch to bring trees to the yarding area rather than a chain to grab hold of logs puts the pulling force at the center of the mass rather than tipped on end to dig into the soil surface.

The log forwarder is little more than a trailer rig equipped with a grapple loader to save the wear and tear of dragging whole stems across the forest floor. Courtesy of *Northern Woodlands* magazine.

Cut-to-length systems allow "short wood" to be moved from the tree stand with a log forwarder. Trailer-type forwarders that are pulled by a tractor can often handle longer logs than an integral unit. Reducing potential damage to remaining trees that otherwise would act as "trail bumpers" is a good thing. Self-releasing snath blocks can be used to get around corners with full-length logs. A little more time and expense makes possible significant nondisturbance.

A return to horse logging would accomplish yet more according to fungal standards. A one-horse

Working with horses in the woods suits the fungi in all respects. Courtesy of *Northern Woodlands* magazine.

skid trail needs to be 5–8 feet (1.5–2 m) wide. Pull logs out with a team of two horses and that skid trail requires 8–12 feet (2–3.5 m) of surface commitment. Bring in the feller buncher and you're looking at as much as 20 feet (6 m) of surefire destruction. So-called equipment efficiency has great ramifications when it comes to preserving mycorrhizal networks.

A typical harvest in northern New England averages 15 cords to the acre: Half of that goes to pulp, 27 percent goes to lumber, 20 percent goes to biomass, and 3 percent goes to pellets and firewood. That's a hefty toll on forest prospects, amplified all the more when the cycle of cutting is as short as every thirty to forty years. Heavy partial cuts like this lead to significant portions of the landscape covered in seedlings and saplings that favor early successional species such as birch and poplar. This deliberate downgrading of the future is made all the worse by whole-tree harvesting, where not even slash gets left behind for fungal banking. Treetops that are chipped and burned for biomass markets are an important source of organic matter to nourish future tree growth. Think "ramially" for a minute about all that cambium containing so many important nutrients to appreciate the full value being expended here. Loss of woody debris is felt all the more in stands with poor soils that are frequently logged for minimal returns.

Industrial forestry does not leave behind a functional forest. Low-impact forestry — what just as well might be called *thoughtful forestry* — offers a far more gratifying scene to contemplate. Nicely shaped trees become all they can be. Smaller trees that are essentially worth pennies on the biomass market are selectively thinned on a canopy basis. Random "wildlife trees" are left for the ages. A well-done harvest will reveal relatively few large stumps in a small area, with barely an opening in the canopy and most adjacent trees left carefully in place. Mixed-age stands are alive with fungi, medicinal plants, and even numerous warblers that eat pests such as the spruce budworm. Species richness (number of species per unit area) will be much higher in a healthy forest than in the sprayed, monoculture, even-age stands created by industrial forestry.

The landowner receives the benefit of ongoing returns over time while allowing the forest to develop in height and volume as well as complexity and quality. We need to consider not just the value of what can be cut today, but also the value of what can be retained. The yields to be gained from low-impact forestry decades into the future will always beat the costs of cutting heavily in the present day. Trees simply don't grow fast enough to support our throwaway society. As my friend Mitch Lansky says, "Calculations involving multi-generational discounting do not send a very pleasant message to our children or grandchildren. We are saying in effect that consumption of aquifers, old-growth, fisheries, or topsoil now is better than availability of these resources to future generations. This is economics without a sense of cultural continuity."[78]

SEEING THE FOREST FOR THE TREES

Healthy forestry creates more jobs as well. Three times the number of loggers will be employed by setting aside mechanized clear-cutting operations. That means more hands-on work, along with green-certified prices for wood and wood products. All to the good! Local economy is not unlike the overarching mycelium of the fungal realm in this sense: No one need be forgotten. Dollars recirculated on a neighborhood scale keep everyone afloat. Similarly, we need to use the higher-quality wood resulting from low-impact forestry practices for value-added products like furniture and locally sawn lumber. Another rich vein to be tapped is that of tourism, for by creating a forest ecosystem attractive to people and wildlife, all sorts of recreational opportunities add yet another component to local livelihoods. Mushroom tours, such as offered in the Pacific Northwest, in turn become a boon for innkeepers and restaurants. The result of a series of good stewardship decisions puts forest communities back on the map.

Closer to my home, the folks at Green Fire Farm in Peacham, Vermont, supplement winter work in the forest with biochar production from the less valuable wood that gets generated in the process of selective thinning and clearing pasture. Talk about economic anastomosis![79] That biochar in turn recharges fungal systems in nearby gardens and orchards, thereby boosting the local food economy. The process involves an Adam retort (kiln) that costs about $4,000 to set up. Each wagonload of wood "cooks" through pyrolysis, about a cord at a time. The finished product gets inoculated with microbes and nutrients, then packed into 2-gallon or 8-gallon bags, which sell for $12 and $40, respectively.

This type of homegrown value-adding is but one example of what an innovative forest entrepreneur can do. A ramial chipping operation set up to use a portion of the slash resulting from thoughtful tree management becomes the basis of a local composting operation. Exploring a range of ways to achieve value with a broad woodlot perspective can be found in *Farming the Woods* by Ken Mudge and Steve Gabriel. High-value crops such as American ginseng, shiitake mushrooms, ramps (wild leeks), maple syrup, and productive nut trees can readily be part of a working forest. Much acclaim goes to agroforestry conversion of open farmland today (see "Agroforestry," page 179), but the truth is that many woodlots are ripe for diversification, too. Those itinerant mushroom pickers working surreptitiously in national forests represent an underground economy that could be replicated in the open by simply seeing beyond timber.

Protecting forest biodiversity and ensuring the sustainable use of forest products comes down to a three-part harmony. Firstly, wild, unmanaged areas ensure that all habitats and species continue, including ones especially sensitive to human meddling. Set-aside reserves are just as important for our spiritual renewal. Secondly, low-impact forestry provides wood products in a sustainable manner while leaving a recognizable and functional forest in place. Finally, we need to temper our collective appetite for wood and paper products. Curtailing consumption ensures that any local reduction in timber harvesting — due to reserves or lower-impact forestry — does not get translated into greater environmental damage elsewhere. Go slow; support hands-on local; pay more but use less.

The forester Alex Shigo signed one of his books to me this way: *Touch Trees!* We connect with Nature by learning about trees and fungi. That's a good starting point for understanding how to truly value the forest in all its giving ways.

Farm

Appreciating soil aggregate formation lies at the heart of fungal-friendly farming. Similarly, increasing soil organic matter reflects a biological emphasis. Still others will speak about "carbon farming" based on our planetary need to capture carbon and sequester it in the soil. It's all one

and the same from the perspective of mycorrhizal fungi. We build on fundamental biological processes when we do things right by the soil . . . and that in turn means we are doing right by our communities and future generations.

Holistic farming practices go hand in hand with higher-quality food. The truth is that the past hundred years of industrial agriculture have done us few favors. The minerals historically derived from rich soil have diminished by more than half. Phytonutrients have declined as a result of bypassing healthy plant metabolism in favor of chemical manipulation. The vast acreage of such annual monocultures as corn, wheat, and soybeans also contribute to soil erosion due to regular soil disturbance. Today, some 70 percent of our calories come from these and other grains grown on about 70 percent of our acreage, both domestically and globally.

We need to shift to perennial polycultures and involve far more people in the growing of food once again. According to Wes Jackson of The Land Institute in Kansas, industrial agriculture is the number one threat to wild biodiversity, and land use is number two as a source of greenhouse gases. His work to discover and promote perennial grain to be planted with legumes, sunflowers, and a few others creates the kind of broad diversity needed to thwart pests and disease pathogens without chemicals. Similarly, farmers are reintroducing tree crops and integrating grazing into ever more diverse operations. Sustaining our place in this world looks all the more doable.

Farming in terms of fungal imperatives defines this turnaround, as I see it. Loss of mycorrhizal fungi comes about in three basic ways: Fallow soil. Excessive tillage. Annual monoculture. Changing the prevalence of these three practices must be the driving motivation of our times. Fungi are

Carbon Perspective

The library for growing healthy on a farm scale just gets better and better. *The Biological Farmer* by Gary Zimmer of Midwestern Bio-Ag has long been my go-to guide for appreciating soil ecology when it comes to crop rotations and mineral fertility. Jerry Brunetti explores the interconnected dynamics of a diversified operation in *The Farm as Ecosystem* like no one else can. And lastly, you have the advantage on a top-notch team like Chelsea Green of passing the ball to a teammate to take convincing shots. The three-pointer, in this case, belongs to Eric Toensmeier for *The Carbon Farming Solution*. Here are the scientific facts and inspiring case studies that point to working with soil biology as the driving force behind carbon sequestration. The time has come to craft a truly sustainable agriculture. *Swish*.

dependent on plants for sustenance and cannot survive for any extended period of time without the presence of living roots. Having a goal to never leave ground bare is as simple as planting cover crops. Tilling — which by itself adversely affects mycorrhizae — often leads to fallow conditions as well. Every grower who finds a combination of ways to rarely break ground to harvest worthy crops is an absolute artist. Microbial brews, grazing livestock, organic no-till, and agroforestry allow us to craft this more enduring agriculture in all sort of ways. Annual crops will always have a place, but the fungal dynamics here need to be underscored. Once harvested, the roots of annuals that are left behind soon die, and any associated mycorrhizal fungi die with them unless new living roots are introduced in the form of another crop within several weeks. Relatively few spores develop with repeated cycles of annuals that include regular intervals of fallow. Root fragments can only carry forward so far without renewal. Interspersing perennial plantings and unmanaged areas into a farm layout helps overcome all this by providing proximity to natural populations from which arbuscular mycorrhizal colonization can be renewed.

The promptings that follow are simply intended to point us in fulfilling directions for keeping biological connection to the fore. Fungal-friendly farming presents entirely new riffs for the post-industrial age ahead.

Farming with Microbes

Stewardship of soil fungi requires *getting there* to begin with and subsequently *staying there* as crops and requisite techniques shift from year to year. Farming decisions affect soil biology to one degree or another. Breaking ground for practical purpose enables agriculture, whether harvesting annual crops or establishing perennial systems. Each and every farmer has the ability to keep soil-life considerations to the fore regardless of scale.

That may well involve tapping into the "biological toolbox" on occasion.

Which is a far cry from reaching for more NPK. Growers who put biology first recognize different priorities. Synthetic fertilization can produce large plants . . . but this suppresses mycorrhiza formation and thus the myriad benefits of fungal connection. Synthetic fertilization does nothing to favor beneficial bacteria or improve soil structure or increase plant species diversity. Nothing. Skeptics who point out how biological products are unnecessary are often the very people buying into the chemical mind-set wholeheartedly. Such will always be the disposition of our world. This quintessential decision to give emphasis to soil biology over soil chemistry comes down to you. The right choices follow from there.

Using mycorrhizal inoculants on a farm scale keys to two factors. The need to heal abused land makes upfront fungal investment fairly obvious. Things become that much more interesting down the road. Improved yields call for plenty of nutritional *oomph*, and the foremost alternative to NPK is the symbiotic partnership that plants establish with mycorrhizal fungi. Monoculture by definition lacks networking advantage. This in turn calls for maintaining a diverse fungal portfolio. Yet crop rotations alone do not necessarily compensate for producing one crop at a time. Cover crop cocktails help tremendously, but don't address every sequence of events in the field. Re-upping enlistment of a fuller array of mycorrhizal species can be worth considering. The silver bullet blarney — *Glomus intraradices* to the rescue — has clearly been shown lacking in terms of yield advantage when compared with a spore blend.[80]

Let's zone in on effective inoculation for significant acreage. Sifting a powder formulation into the seed box of planting equipment works well with hairy-textured seeds such as wheat, barley, or oats. Seed adhesion ensures inoculum proximity to the germinating seed and prevents a buildup of excess inoculum in the seed box that might clog the grain drill. Smooth-surfaced seeds, on the other hand, work better with a liquid formulation. Spores adhere readily to corn, beans, and alfalfa when inoculum suspensions dry in place. Some suppliers will coat seeds with appropriate inoculum, if you request that service. Seed can be treated directly on the farm with an electric cement mixer or by lightly spraying the seed as it passes up through the filling auger to the seed box.[81] Legume seeds typically are treated with rhizobacteria to fix nitrogen, which can be premixed with mycorrhizal inoculum in order to get both applied in one fell swoop.

Nor should biological farmers forget the rest of the team. Farming with microbes calls for ingenuity in covering more ground, but the advantages of ever-targeted diversity remain one and the same. This quick summation of "applied biology" explains what's worth doing to further facilitate mycorrhizal association for the crop at hand.

—————

Aerated compost tea (ACT). The diverse range of organisms in good compost responds with zeal when conditions are provided for rapid multiplication. Introducing air by means of a bubbler and microbial foods favors those species that in turn proliferate in a healthy soil food web. Lessons in light microscopy are in order when using compost tea for competitive colonization purposes to ensure that the brewing process has a somewhat fungal edge. Tea microbes are actively metabolizing and thus ready to engage immediately on the leaf surface. Aerated compost tea must be used immediately following the twenty-four- to forty-eight-hour brew cycle.

Effective microbes. Facultative organisms have the ability to function in both anaerobic and aerobic environments, which lend them to a passive brewing process. Photosynthetic bacteria, lactic acid bacteria, and specified yeasts are the primary organism groupings in commercial mother cultures. Ground applications boost decomposition of surface organic matter, whereas canopy applications play assorted functions in making foliar feeding biologically effective. Allow a day or two for these microbes to awaken from the anaerobic state in competitive colonization scenarios.[82]

Indigenous brews. What's now offered commercially as effective microbe formulations was originally a locally inspired affair. Just ask anyone in the natural farming movements in Japan or Korea. The Permaculture Research Institute has posted an impressive recipe on "How to Prepare a Beneficial Microorganism Mixture" on its website.[83] Similarly, the microbes associated with freshly harvested herbs, whole milk, and worm castings in a fermented plant extract contribute native diversity to these homemade brews. The activated microbe component used in this approach to making calcium- and silica-rich extracts (see "Fermented Plant Extracts," page 84) helps to jump-start the fermentation as well as bring photosynthetic bacteria into a fuller state of readiness. Biological methods don't get more economical than this.

Liquid compost extracts (LCE). Compost extract technology is geared toward broad-scale application. Aerated compost teas can't touch the volume and extended life that liquid compost extracts offer to farmers with significant acreage. Organisms are extracted directly from good compost (or worm castings) and held stable with no further aeration or feeding. A higher degree of biodiversity can be found in LCEs than ACTs, with most microbes left in a dormant state, allowing for ten to fourteen days of holding time. Liquid compost extracts act as a catalyst to enhance food web activity in soils across the board.

Beneficial bacteria. Nitrogen-fixing bacteria paved the way in making most growers aware that "small critter inoculation" offers apt purpose in defined scenarios. *Pseudomonas* bacteria are easily wiped out by glyphosate early on, for instance, and yet certain of these species are among the best detoxifiers known. Time for staging a comeback, methinks. Growth-enhancing bacteria occur naturally in soils, yet not always in high enough numbers to have a dramatic effect on the intended crop. Bacterial inoculants abet the work of mycorrhizal fungi as well (see "Bacterial Metabolites," page 79) by making immobile nutrients available for hyphal uptake and reducing compaction for roots and mycelium alike.

Biostimulants. The use of seaweed extracts to boost microbial activity and plant growth should be part of such applications. The cytokinins found in kelp function at very low rates of application — as little as 8 ounces dry weight to the acre — especially compared with spreading compost by the tons. This array of plant growth hormones stimulates plant metabolism, thereby enhancing plant root function. Better growth in turn benefits the biology, leading to improved soil structure.

Moving agricultural soils a few degrees toward fungal dominance is the right goal. Surface decomposition of organic matter can be assisted with these microbial boosts. Healthier plants in turn have a greater photosynthetic rate, resulting in more carbon currency flowing to soils. The crescendo builds as yields become more profitable, due in part to no longer having to rely on chemicals.

Meaningful crop production can be accomplished with biology instead. Organism-based solutions can be homegrown or produced on a regional scale, which in turn opens up business opportunities in rural areas. Spraying microbial inoculants in field situations is part of the dance that makes farming so enjoyable . . . and, truth be known, an integral part of the agriculture that will deliver real-time food security for us all.

Grazing

The root systems of grasses respond differently when grazed by livestock, according to stage of growth and the frequency of the grazing. Grasses that are nibbled to the ground on a somewhat regular basis are depleted, with straggly roots going to no particular depth. Grasses that reach the two-leaf stage in a casually run rotational grazing system have roots that reach nearly twice as far, but lack mass at depth. Grasses grazed no more than twice in a season at the three-leaf stage (just prior to initiating seed) have roots with the greatest depth of all and significant mass throughout the soil profile. More roots represents greater fungal opportunity going ever-further down.

Grasses have the oldest cells at the leaf tip, unlike woodsy plants and broadleaf forbs, which grow with the youngest cells at the shoot tips.

Therefore, grazing animals can remove the upper leaves of grass without stopping the growth of new leaf buds (tillers) located along the shoot close to the ground. These remain below the reach of the grazing animal as long as the shoot is in the vegetative, or nonflowering, phase. Well-timed grazing that removes the upper portion of a pasture results in a 30–45 percent increase in herbage production. More green, more photosynthesis, more carbon. All to the good, say the fungi.

Livestock tugging on plants stimulates greater rhizosphere activity as well. Mowing results in a clean slice that closes readily. Herbivore damage, however, requires more proteins to close off the tear, thereby releasing more carbon to the roots as a result of shoot mobilization of amino acids. A portion of roots will be shed outright as well, as the plant rebalances its shoot-root ratio. Livestock maintain the soil food web through trampling as well. Treading on what remains after an intense round of grazing results in the surface of the soil being covered. This protects soil microorganisms while the system takes a breather, preventing evaporation and moderating soil temperature at the same time. Throw in the occasional cow pie and subsequent breakdown of carbon-rich organic matter — coupled with no soil disturbance whatsoever other than this kneading of hooves — and it's pretty obvious that proper grazing is a downright fungal act.

The excitement of animals brought to fresh grazing is unmistakable. Every bite turns out to be the absolute best. The fungal network below must sense much the same given what's about to happen. Higher densities of AM fungi are found in healthy perennial grasslands than in any other plant community. All that liquid carbon flowing downward builds soils in "grass fashion" precisely because animals eat grass at the right stage.

What's been deemed *holistic planned grazing* is a crucial means of sequestering carbon in the soil. The worst paddocks in the world can be regenerated back to health by jump-starting this fungal cycle into motion through high-density grazing. Pioneer species can be invigorated with a onetime application of compost. Grazing animals for short bursts will start to build the diversity of grassland plants. More and more plants retain more and more water. The carrying capacity of the land with regard to stocking rates increases twofold, then fourfold. Animal husbandry becomes more profitable as grass-fed meat becomes more widely available. Two-thirds of landmass is grass, and if

only given better management, we would see big gains fast in terms of planetary recovery.

This is why books like *Cows Save the Planet* by Judith Schwartz and *The Art and Science of Grazing* by Sarah Flack are so important. Innovative farming practices such as introducing summer-active perennial grasses into pastures, rotational (originally deemed *rational*) grazing, and pasture cropping cereal grains have profound impact on soil carbon. Cover cropping of production fields on diversified farms ties in here as well by extending the grazing window while pastures recover. Ranching becomes regenerative via biomimicry through this holistic grazing approach. Everything comes

Vast amounts of soil carbon can be sequestered in pasture ecosystems when grazing is done right.

together for the greater good of the herd and the soil . . . and those ever-appreciative fungi.

ORGANIC NO-TILL

Nearly seventy-five years ago in the first paragraph of the landmark book *Plowman's Folly*, Edward Faulkner said, "The truth is that no one has ever advanced a scientific reason for plowing." Looking out over freshly turned earth in a field prepped for planting in spring is a joyful sight . . . but let's consider this from a fungal perspective.

Turning the soil with heavy equipment results in a broken structure lying atop a heavily

The lushness of a rye-vetch cover can be turned into effective organic mulch in no-till systems.

compressed plowpan (the undisturbed layer that the plow or tiller doesn't reach). The advent of no-till farming in conventional circles over the last several decades improved things compaction-wise, reduced erosion, and indeed preserves a tad of mycorrhizal happenstance. That last bit stands revealed by slightly improved levels of glomalin content in no-tilled ground compared with plowed ground.[84] Use of Roundup in what's called a "burndown application" in the spring kills off any weeds in the field prior to planting. And therein lies the bugaboo. The active ingredient in Monsanto's badass herbicide, glyphosate, kills the soil microbiome just as an antibiotic kills the microbiome in one's gut.[85] Reducing the physical onslaught on soil life by means of a chemical fallow has its proponents, but still comes up far short of the mark of what could be.

Controlling weeds with plant competition offers the better solution. Bulking up on organic matter through cover cropping provides an ongoing nutrient base by which all microbes thrive. Living roots carry forward mycorrhizal populations. The question begging to be answered was how could organic growers work with a continuous living system and yet avoid constant tillage in order to plant the next crop?

A guy looks around and notices things. Jeff Moyer, longtime farm manager at the Rodale Institute, observed an interesting edge effect alongside freshly turned fields. Where the overwintered rye-vetch cover crop had been run over with tractor tires, stalks essentially got "crimped" in one or more places and subsequently died without detaching from soil. The thick mulch left in place suppressed weed expectations. An idea was born. Jeff spent the next several seasons perfecting a roller-crimper, to be used on the front end of the tractor, with pressing weight adjusted by water fill (the roller unit can

Jeff Moyer deserves great credit for refining the crimp roller, which has made organic no-till possible on a farm scale. Courtesy of the Rodale Institute.

be topped off or not to meet varying conditions in the field). A seed drill attached on the back end of the tractor makes possible planting into the resulting mulch mat with roots and fungi still intact.

The four basic steps of "organic no-till" can be summarized thusly:

protect soil with winter cover
knock down at maturity
drill seed into residues
harvest crop, then disk and begin again

Success lies in the details. The chosen cover crops need to be in the flowering or early grain-heading stage for crimping to be fully effective. Breeding efforts to develop vetch strains with

earlier maturity have allowed growers to get the planned crop in the ground sooner. Planting into thick residues requires adaptations to traditional seed drills. And finally, we should recognize what is meant exactly by the word *no*.

Zero tillage is by all means the fungal ideal. Yet the terms *zero-till* or *no-till* can be somewhat misleading. What organic farmers are really trying to achieve is "reduced tillage" agriculture . . . greatly reduced tillage, in fact, when looked at over a multitude of growing seasons. Crop rotations ultimately tie to markets as much as soil health. (We'll leave the fallacy of "continuous corn" to the chemical camp.) A diversified organic farm may have this all worked out with grain crops and soybeans and the like, but there's still great benefit to a down year or two for livestock to forage. Similarly, market gardeners working with direct-seeded crops such as carrots and microgreens need a shallowly tilled bed surface for ideal germination. Going about *organic no-till* doesn't rule out biological compromise in certain years. Doing things wisely in the context of making a living means using different tools at different times to keep fungal connections pulsing as best as we're able.

Cover Crop Cocktails

The greater the diversity of plants, the more resilient the fungal network over time. Thoughtful mixtures of cover crops tie right in with carrying forward mycorrhizal prospects. What's applicable to the farm suits equally in the garden . . . only we scale up for serious acreage.

The cocktail approach to cover cropping involves using combinations of plants — anywhere from half a dozen to twenty or more species in a single mix — to achieve multiple soil-health goals. Cover crop mixtures can be interplanted between taller growing production crops such as corn. Just as often, the cover crops are planted after a summer-harvested crop such as wheat, directly into the stubble. The overriding rule of thumb is to never leave ground bare. Year-round cover-cropping programs fit nicely with a livestock component. Animals benefit from grazing the top third of the aboveground biomass at the peak of lushness, trampling the rest, and leaving behind intact root systems through the winter months. Springtime brings a series of options, depending on rotation plans for the next cash crop.

Designing a good mix hinges on several factors. Time of planting determines whether to prioritize warm- or cool-season varieties. Cereal grasses, broadleaf plants, and legumes make up the three broad categories to weave together. The Cover Crop Chart put together at the Northern Great Plains Research Laboratory in North Dakota has been modeled on the periodic table of elements to correlate the many options. This represents a compendium of information from multiple sources throughout the United States and Canada, making the chart applicable across the continent. The interactive version on the USDA website (see "Soil Health Resources," page 201) allows users to click on the name of individual crop species for additional information on growth cycle, relative water use, plant architecture, seeding depth, forage quality, pollination characteristics, and nutrient cycling.

The real fun begins as you see the resulting ecosystem develop. Legumes not only fix nitrogen but provide nectaries for pollinators and beneficial insects. Grass root systems greatly increase soil aeration, reaching depth much quicker than early growth up above would seem to reflect. Insistent taproots literally drill through hardpan to break up soil compaction. Fast-growing broadleaf species will tone down a move on the part of aggressive weeds. Brassicas scavenge for nutrients and provide an excellent home for earthworms. All told, it's the consortium of roots that will bring about significant increases in soil organic matter.[86] Commercial mixes are available from numerous companies. Throw in some annual wildflowers and aromatic herbs for good measure if a forage plan will likely extend through a full growing season.

Mycorrhizal inoculation keys to the cropping history of the field in question. Years of biological management will have kept the fungal dynamic to the fore — no worries, plant away — while hefty use of herbicide and other synthetics in the not-so-distant past makes the call for action paramount. The inoculation opportunity presented by a broad

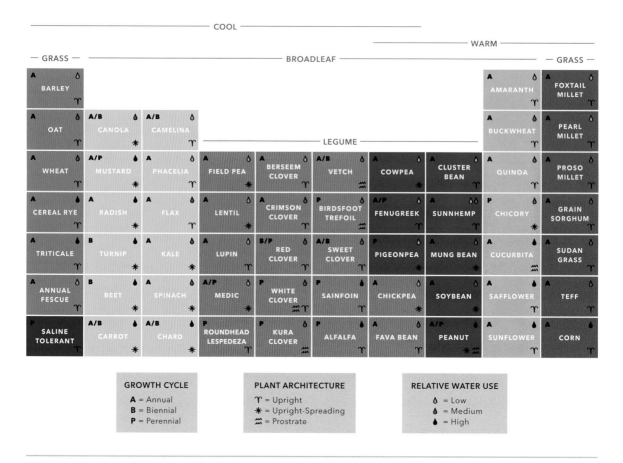

USDA Cover Crop Chart. Courtesy of Mark Liebig, Northern Great Plains Research Laboratory, USDA-ARS.

diversity of cover crops is about as good as it gets when investing in multiple species of arbuscular fungi to up underground diversity. Even a nonmycorrhizal species or two in the mix can't mess with everybody else finding a home.

An often overlooked part of mycorrhizal carryover is "roots in the ground" on both sides of the dormant season. During warm spells in late fall and again in early spring, fungal spores will try to germinate and hyphae from the previous crop will reach out from root fragments . . . but these can exhaust carbohydrate and lipid reserves if no active plant hosts can be found. Farmers opting to leave ground bare over the winter months unwittingly

allow the fungal system to become much weaker by the time soil temps rise enough to put in a warm-season cash crop. Cover crop mixes designed to overwinter, or even "living stubble" left after fall grazing, will maintain healthy AM fungal populations for the next go-round.

Seed-drilling multiple cover crops into grazed or occasionally rolled residue of the previous cover crop opens the door to whopping increases in organic matter. Equipment capability has significant bearing on preserving both what was and what will be when talking about acreage. (*Hectarage* for all you metric converts!) Mixing different-sized seeds with precision and then placing

The duoseeder lays down two rows of cover crop seed within each corn aisle to create a diverse biological understory. *Courtesy of Dawn Biologic.*

radish-sized to bean-sized seeds at a workable depth for all requires ingenuity. Approximately forty seeds per square foot work well. Bringing in the herd to graze again in late spring, if continuing on to round two of a long-term soil-building plan, brings the "hoof seed drill" into play. Growers establishing trees on contours have even used seed balls à la Masanobu Fukuoka between swale plantings to bring equipment-worked ground back into shape. Mycorrhizal inoculant can either be sprayed as a liquid formulation onto the seed or, if the seeder has a second legume box, applied directly into each planting furrow in powdered form with the clover seed.

Agroforestry

Agriculture is the production of not only food but also fuel, fodder, fiber, and medicine. Just as diversification of the woodlot opens up entirely new horizons, bringing trees and permaculture principles to open cropland makes possible a fungally integrated way to farm.

Agroforestry in its most basic form involves animals grazing mixed forages between rows of trees. Trees are planted on or near contour, often with some element of keyline subsoil plowing and additional water retention through the use of swales. Alley cropping of annuals can fit in on a rotational basis in the early years. Livestock does well on silvopasture (trees on pasture) in part due to the increased browse diversity of understory plantings. Windbreaks and riparian buffers fall into place by simply planting more trees.

J. Russell Smith came out with his classic work, *Tree Crops: A Permanent Agriculture*, back in 1929. He saw the usual problems with annual agriculture and realized that a multitude of trees could solve erosion and nutrient-depletion concerns quite handily. (Let's put this man on our list of fungal seers!) The species outlined in this book are basically the stalwarts of agroforestry in temperate zones today: assorted oak and sweet chestnut, hickory and pecan; apple and pear and cherry and mulberry; hazel crosses and mesquite, persimmon and pawpaw, and even carob in the California foothills. Throw in black walnut for its esteemed timber value, along with black locust for forage and post wood. Particular fast-growing tree species are continually chopped as ramial mulch for the nut and fruit trees. Nitrogen-fixing trees become especially valuable in subtropical and tropical permacultures (where the choice of fruits literally goes bananas in turn). All sorts of fruiting bushes can be tucked into understory niches along with other multipurpose herbaceous species. Mycorrhizal inoculation of planted trees is extremely important to jump-start conventional ag ground and introduce far more fungal diversity in the form of both AM and EM spores.

The focus of all this varied planting is to maximize output while reducing inputs through an exchange of ecological services. No two agroforestry operations are alike, as each system is designed according to the individual specifications of the farm and farmer. Interest in particular livestock helps determine things like tree row spacing and suitable varieties.

Versaland in Johnson County, Iowa, offers a wide-sweeping view of agroforestry in the heartland. Grant Schultz bought a soil-weary farm and proceeded to prepare for a new way of farming. Step one was to subsoil this rolling acreage on contour using a Yeoman's noninversion subsoil ripper. Swales were used on steeper ground, but for the most part were unnecessary. Some thirty thousand trees were planted with the help of many dedicated volunteers. A rich mix of legumes and other cover crops created lush grazing between tree rows. The apple emphasis draws my attention, from heirloom russets to classic cider apples, but there are plenty of nuts and everything else planted

Tree rows on contour have been launched at Versaland in Iowa with the help of hardworking volunteers. Photo by Grant Schultz.

as well. Honeyberries, currants, and gooseberries are typical understory shrubs. Nitrogen comes from leguminous hay mulch at the base of the trees and manure being spread fresh out of the rear ends of grazing animals. The alley width between tree rows is 30 feet (9 m), which works well with the sheep, pigs, chickens, ducks, geese, and turkeys steered throughout the farm every day. Beef cattle are tougher to manage in narrow alleys and so are kept to the edges of the tree plantings. All together that may sound overwhelming, but the course of "tree time" allows for integrated markets to be developed in conjunction with getting everything in and growing.

What I find especially exciting with integrated farming approaches is the reclamation of degraded soils across the globe. Agroforestry efforts in Niger, for example, have resulted in two hundred million trees being planted on over five million hectares of farmland. This impacts many people by improving soil and creating resilience against climate change. Fazenda da Toca is an organic family farm in São Paulo, Brazil. Agroforestry techniques are applied throughout the farm, with 2,200 hectares under integrated cultivation, earning livelihoods for some 140 people from egg, dairy, and fruit production. India has a full-blown governmental policy with funding incentives to increase tree cover through

Agroforestry projects across the globe are coupling trees and permaculture techniques to create viable farming opportunities for many people. Courtesy of Trees for the Future.

agroforestry. The total area under agroforestry in the European Union is estimated to be about 24 million hectares (which is equal to about 14 percent of the utilized agricultural area). The nonprofit Trees for the Future finances numerous tree-planting projects in Cameroon, Kenya, Senegal, Uganda, and Tanzania. Families learn to plant trees with traditional crops and open-pollinated staples to grow a wider diversity of food, and thereby build a sustainable future. Hundreds of thousands of acres of soil have been revitalized in sub-Saharan Africa . . . thereby changing people's lives forever.

Agroforestry systems ultimately become much like an open savanna ecosystem as the trees mature. The mechanization required in the early years to subsoil and plant cover crops will be less needed over time as long as grazing remains an active component. Interesting shifts begin as canopy shading becomes a more dominant factor. You can see where all this is heading fungally. Many plants with multiple endomycorrhizal and ectomycorrhizal affiliations in such close proximity create an extremely diverse mycelial support system. Water distribution, check. Nutrient balance, check. Defense messaging, check. The resiliency everyone touts through agroforestry comes about primarily because a rock 'n' roll mycorrhizal network develops on richly integrated farmland that includes trees and grazing.

CHAPTER 7

Edible Mycorrhizal Mushrooms

Call this the symbiosis benefit especially for us. Ectomycorrhizal species evolved from saprotrophic fungi with fruiting bodies. What an entirely delectable trait to pass along!

I claim few talents as a stalker of edible mycorrhizal mushrooms. The occasional morel on abandoned apple ground, the golden glow of chanterelles easily found — knowing these gifts are out there is happiness in itself. Yet now I'm motivated to spend even more time in the woods, thanks to inspiring friends with tried-and-true mushroom abilities. What follows is an introduction to an array of edible wonders. This is by no means a mushroom identification guide . . . you'll need one or more of the titles suggested at the end of this chapter to be sure. We begin with the unquestionable tie between mycorrhizal mycelia and the root systems of certain tree species.

Identifying trees is part of the skill set needed to know where to best look for these mushrooms.

Some mycorrhizal fungi form partnerships specifically with manzanitas and madrones, for instance, along the Redwood Coast. Other species are found only with alders, or with certain pines, or with beech. On the other hand, the same tree can have numerous mycorrhizal affiliations and thus offer prospects for different fungi at different points in the year. According to David Arora, author of *Mushrooms Demystified*, "[t]here are well over 1,000 kinds of mushrooms known to form mycorrhiza with Douglas fir, and the great Douglas fir forests of the Pacific Northwest are among the best fungal foraging grounds in the world."[1] Mycelia extend further than the canopy stretch of a single tree, so cast your eyes broadly across the forest floor. The "tree viewpoint" becomes more about recognizing the right sort of forest ecosystem where desirable mushrooms will be found.

My better chanterelle finds have come in mixed stands that include balsam fir; the same place where others report finding boletes. Mature hemlocks along streambeds are noted for being a good place to look for dozens of species, including matsutake. The oak family supports many species including king boletes, chanterelles, and black trumpets. Birches are a well-known source of chanterelles, hedgehogs, and the more disconcerting webcap.

Let's zone in all the more as to the right sorts of places to look. Mixed stands of hardwoods and conifers provide a variety of habitats for mycorrhizal collaboration. Mature trees are more likely to offer a prize than a younger, successional stand. Look for sphagnum moss, in part because mosses reflect nondisturbance and thus fungal continuity. Well-shaded embankments along back roads and trails are prime mushroom ground. Any microclimate that holds moisture tends to be a good fungal bet.

Establishing mycorrhizal mushrooms at one's own place requires patience. The first step involves getting mycelium growing on the right sorts of trees . . . and then you wait and wait some more. Years may pass before a single fruiting body shows itself. The "lottery approach" involves distributing spores from forest finds on promising tree sites. Overly mature specimens or stem butts serve well

Yellow morels are one of the most readily recognized of all the edible mushrooms. Photo by Mary Smiley, Creative Commons.

here. Scratch these slightly into the ground to get closer to the proximity of actual roots. The odds may be slim that the new species takes up residence (by which point you will have forgotten all about it anyway!), but those spores will indeed germinate. Only time will tell if an introduced species can wriggle its way into an established fungal scene. Transplanting saplings from proven mushroom ground brings with it the expectation that those root systems will carry the same mycorrhizal connection along. This holds slighter better promise, only it may be a decade or more before those trees come into their own and that first mushroom says *here I am*. Similarly, you can make root dip inoculum by crushing the spore-bearing surfaces of wild mushrooms (dried and held over from the previous fall) into water and then immersing the roots of appropriately matched nursery stock prior to planting those seedlings in early spring. Chances of success rise if such root systems have no dominant ectomycorrhizal affiliation to start.

Morels are among the better mycorrhizal candidates for establishing on home ground. These mushroom species work in a broad range of host situations thanks to a ready ability to adapt. Some form mycorrhizal associations directly, while others express a modus operandi that's more saprotrophic. Some believe that morels are disrupted out of a healthy symbiosis when a tree dies for whatever reason . . . causing the fungus to withdraw from the roots, form sclerotia (compact masses of hardened mycelium, with food reserves), which then go into fruiting mode separate from the tree in order to sporulate and find new hosts. This dual nature goes a long way toward explaining the settings where these delectable mushrooms can be found.

Savvy morel hunters know to look near ash, oak, and wild cherry trees in forests with moderately well-drained soils. Cottonwoods and tulip poplars are good indicators of yellow morels in the vicinity. Mushroom hunters from Tennessee claim to find these just as often under red cedar. Mass fruitings of black morels follow forest fires in the coniferous forests of the Intermountain West. Morels continue to feed off the root systems of dead elms and apple trees in abandoned orchards as true saprobes. Morels are even found in gravelly roads and streambeds on a go-figure basis.

Conditions are right early in the spring as the ground is warming. Morels will first appear on south-facing slopes in fairly open areas. As the season progresses, the place to be is deeper in the woods on the north-facing slopes. The season begins when trillium flowers bloom and redbud trees burst forth with color. Here in northern New England, morels peak when apple and lilac bloom overlaps. Morel season can be very short in southern regions when it gets very hot and dry early, while the cool, wet weather often experienced further north is conducive for stretching out the harvest window to a full month long.

You'll find strategic tips for cultivating morels in *Organic Mushroom Farming and Mycoremediation* by Tradd Cotter. Regional strains have a strong affiliation, not only for specific tree types, but importantly for the soil microbe community found in each place. Positioning a nonnutritive layer between the spawn and wood chip substrata is integral to getting morels to pop forth the next year. Native soil — from where the morels being cultivated were found, no less — must be included in the propagation bed to ensure the presence of the right bacterial associates known to that particular morel strain.

Morels are among what mycologists call the Foolproof Four, a list that includes morels, puffballs, sulfur shelf mushrooms, and shaggy manes. Each of these mushrooms has a distinctive look

The bright coloration of chanterelles cheers up all sorts of woodlands in late summer. Photo by David Spahr.

The black trumpet rates as one of the best-tasting mycorrhizal mushrooms. Photo by Tammy Sweet.

that's easily differentiated from other, poisonous varieties. The takeaway here is you need to know all the clues when it comes to identifying any mushroom as edible. Morels have a distinctive conical shape, with a series of cups arranged haphazardly on the head, atop a hollow stem. Yet there are false morels just as there are false prophets . . . so be forewarned.

Chanterelles are delicious, orange-yellow, almost trumpet-shaped mushrooms, said to emit a pleasant apricoty smell. These are found in many kinds of woodland and are fairly easy to spot at that because of their glowing vibrancy from head to toe. The fruiting body can persist as long as two to three weeks, so leave the smaller ones to expand in size for later picking. Look to mossy areas where the right tree associates grow

nearby. This can be anything from oak and beech to hemlock and pines, whether found in uplands or lowlands. David Spahr in Maine points out that the edge of a dirt road or trail can be especially promising, perhaps because the compacted earth to be found in such places causes the mycelium to react by fruiting heavily. July through September marks the months to be on the lookout for chanterelles, especially in those years with consistent summer rains.

One of the identifying features of a true chanterelle is the "false gills" on the underside of the cap. These are not individual structures that sit separate from one another but rather are mere folds in the undersurface of the fruiting body. Assessing *true* or *false* gills is especially important if one wants to eat chanterelles, since the poisonous Jack O'Lantern

Hedgehog mushrooms have spinelike bristles on the underside of the cap. Photo by Greg Marley.

does the same after a heavy rain in a wet summer) are bonanza territory for trumpet lovers. The trick to finding these naturally camouflaged mushrooms is to stand directly overhead, looking down rather than letting your eye rove across the landscape. Some say that looking for black trumpets is like looking for holes in the ground, which isn't far from the truth.

Hedgehog mushrooms are known as *tooth fungi*, so named because of spinelike bristles on the undersides of their caps. The sweet, nutty taste and crunchy texture give these mushrooms a high rating in the kitchen. The cap starts off lightly colored, picking up a yellow to light-orange to even brown hue by maturity. Hedgehogs often develop an irregular shape, especially those crowded in closely with adjacent fruiting bodies. This basidiomycete has no poisonous lookalikes and will rarely be bothered by slugs or other insects. Hedgehogs associate regularly with Scots pine and hemlock, and occasionally with beech and yellow birch. Tooth fungi can be found growing on bare ground, such as eroded river gravels, likely assisting trees to stretch roots into new territory. These mushrooms grow in profusion in the leaf litter of both coniferous and deciduous forests. Fruiting occurs from midsummer through the autumn months.

Concealed under duff on the forest floor, matsutake mushrooms come on strong by September. These highly sought-after mycorrhizal mushrooms grow across all the northern temperate zones. Species on each continent are classified separately, but in truth they are much the same. *Tricholoma magnivelare* is found in the coniferous forests of the Pacific Northwest, is associated with low-growing hardwoods in California and parts of Oregon, and is generally found in jack pine and hemlock forests in the Northeast. The North American variant is typically called "white matsutake," as it does not

mushroom is a cluster-growing lookalike with true, rather than false, gills.

The ability to hide — and hide quite well at that — belongs to the black trumpet grouping of mycorrhizal mushrooms. All species are not only edible but among the very best in flavor. Black trumpets will be called the *black chanterelle* or *horn of plenty* in some regions. These funnel-shaped mushrooms can be tan to light gray early on, darkening to dark gray or black at full maturity. The fruiting bodies range in height from 1–6 inches (2.5–15 cm) and up to 3 inches (7.5 cm) across. The flesh is thinner than most, and extremely fragrant. Small clusters are typically found in mixed woods where beeches and oaks grow, especially shady and damp locations. Those so-called spring washes where snowmelt runs off hillsides as a visible stream (and

Matsutake grows primarily under conifers in north-ern and mountainous regions. Photo by Tomomarusan, Creative Commons.

This pristine king bolete clearly shows the fine stem reticulation so characteristic of this species. Photo by Greg Marley.

feature the brown coloration of the Asian strain. The odor of the matsutake is its most distinctive yet hard-to-characterize feature. David Arora described it as "a provocative compromise between red hots and dirty socks,"[2] and so it may be. The fungus often grows in fruiting arcs through the soil as root outreach extends further each year. Matsutake favor what are known as podzol soils. This gray clayey, somewhat sandy earth is the third layer below a needle layer and a thin humus layer, followed by subsoil.

The king bolete is a delicious, meaty mushroom that grows worldwide. Its many names include porcini and pennybun, comprised of several closely related species with similar looks, habitat, and flavor. These stately mushrooms feature a thick cap, with a distinctive spongelike texture on the underside. Their clubbed stem often appears to be covered in fine webbing (reticulation). Boletes have a strong affinity for spruces and, depending on how the species is defined, other conifers and even hardwoods. The choicest specimens can be served raw, thinly sliced with lemon juice and oil . . . though more typically boletes will be sau-téed in their own juices. These prized mushrooms have a very short life cycle, harvestable for just a few days before turning into maggot hotels and soon after that into a puddle of black slime. The best things in life are often ephemeral.

Edible, of course, doesn't necessarily mean a thing is worth eating. The impressive-looking *Cortinarius armillatus* is mycorrhizal with white birch, and is a common late-summer and fall mushroom in northern forests. The stem has a swollen base, with a series of two to four bright, orange "bracelets" encircling the stem. Its genus name is a reference to the partial veil or cortina (meaning 'a curtain') that covers the true gills when caps are immature. Other species in this family associate with big-toothed aspen, beech, balsam fir, white pine, red pine, assorted spruces, and ericaceous shrubs. The reddish variants of webcaps contain orellanine and are known to be deadly poisonous.[3] The question is, why bother to eat one with a bland

The magnificent gills of *Cortinarius armillatus* are but one indicator of an edible mushroom with some very poisonous cousins. Photo by David Spahr.

The highly rated black truffle often requires a well-trained dog nose to detect these buried treasures. Courtesy of Australian Truffle Traders.

radish odor when its not-so-nice cousins of similar appearance could finish you off?

The fruiting bodies of truffles grow underground and are sought out by gourmet chefs the world over. These mycorrhizal treasures are harvested traditionally in Europe with the aid of female pigs, which are able to detect the strong smell of ripe truffles beneath the surface of the ground, and then become excited.[4] This plan comes with risk, however, because of a pig's natural tendency to eat whatever is remotely edible. Enter the truffle dog, willing to exchange that smelly ol' soil ball for a treat and a pat on the head.

The flavor of the truffle is directly related to its aroma. This reaches peak essence only after the spores are mature enough for release, so truffles must be collected at the proper time or they will have little taste. The resulting flavor can be described as garlicky with cheese overtones . . . yet there are species that come across tasting more like tar with a delicate hint of methane. The point being it matters which tuber fungi of some two hundred identified species are the truffles of choice. The Italian white truffle comes with the highest regard, followed by the Périgord black truffle of France. The Burgundy truffle was held in great esteem in Victorian times. Here in North America, the Oregon white truffle is making waves of its own.

Appropriate tree seedlings such as oaks and filberts are inoculated with truffle spores and then transplanted out to the proper environment (usually a barren, rock-strewn calcitic soil) to establish a truffière. This was a hit-or-miss proposition for the longest time until recent understanding about keeping other ectomycorrhizal fungi out of the picture came to be. It takes nearly a decade before the first truffle begins to grow, with a productive life for the bearing tree expected to last about

The "lobster mushroom" is actually a separate fungus that has parasitized another mycorrhizal fruiting body. Photo by James Nowak.

Laccaria laccata are highly variable mushrooms that start off displaying vibrant tones but then become colorless and drab. Photo by Zonda Grattus, Creative Commons.

The Red Hot Milk Cap tastes mild initially, gradually becomes extremely hot, and then burns you to a crisp. Photo by Zonda Grattus, Creative Commons.

twenty to thirty years. Truffle horticulture is fascinating. Simple things like planting density and tree arrangement hold the key to prolonged production, as the soil surface needs good sun exposure to favor the *Tuber* genus over other competing fungi.

The truffle has been described as the fungus worth more than gold. I may never taste a finely prepared truffle dish, just as I may never open a thirty-year-old bottle of Bordeaux. Instead, I got to read a great book on truffles while drinking my own cider, and our dog came by and I gave her a pat on the head. Reality is what we make of it.

The "lobster mushroom" is the result of another fungus parasitizing either a *Russula brevipes* or *Lactarius piperatus* mycorrhizal mushroom. This produces a vivid orange covering over the mushroom, not unlike that of a cooked lobster, the surface of which is rather hard and dotted with tiny pimples. Eventually, the fungus even transforms the shape of the host mushroom, twisting it into odd contortions. These can be found under a variety of trees, with hemlock and Douglas fir as especially good candidates. Most will be found in the fall months, but "lobsters" can appear undergoing transformation as early as July. These mushrooms are very noticeable: Nothing else looks remotely like fungi taken to the second power.

Laccaria laccata, commonly called waxy laccaria, is a white-spored species of small edible mushroom found throughout North America and Europe. This highly variable mushroom can be deceiving, eventually looking quite washed out, even drab, though when younger it often assumes red, pinkish-brown, and orange tones. Some consider this species to be a "mushroom weed" because of its ubiquity and plain stature. Waxy laccaria taste mild; however, before tasting, you need to distinguish it from other potentially lethal, small brown mushrooms also found in mixed forest ecosystems.

Lactarius rufus is more commonly known as the Red Hot Milk Cap. Its delayed-action effect masks an extremely hot (maybe the hottest) mushroom. These can be used as a condiment only after processing with vinegar or salt. The flesh exudes a milky latex fluid when the tissues are damaged. This entire genus of mycorrhizal fungi is dependent on the occurrence of specific host plants, and in this case, that would be pine trees followed by birch or spruce. These mushrooms appear from late spring to late autumn in northern zones.

We're ending this appropriately enough with a visual delicacy found between the start of fall colors and the onset of snow. Purple-gilled laccaria are

Laccaria ochropurpurea mushrooms are a delight to find in late fall to wrap up the wild mushroom harvest. Photo by David Spahr.

Myco Gardening

The edible fungal realm includes many saprotrophic species that break down organic matter. You can read about inoculating wood chips, freshly cut logs, stumps, straw bales, and even cardboard with chosen spawn in mushroom farming books. Savor shiitakes, reishi, and oysters grown on your own land! Here's one to grow in the garden in

Wine cap mushrooms have a strong affinity for corn planted in stubble-rich ground. Photo by Will Brown.

mycorrhizal with hardwoods and conifers, growing alone or scattered about in pairs or grouped gregariously. These mushrooms are especially partial to oak and beech, but will also frequently be found in young stands of white pine. This species is easy to identify with its whitish-cream cap and dark-purple to light-purple gills underneath and a hard (not quite chewable), fibrous stem. *Laccaria ochropupurea* have a pleasant mushroomy flavor, just right for adding to a hearty bean soup or association with the roots of organic sweet corn and scattered about the orchard.

Stropharia rugoso-annulata is the surest and most productive fungus for myco gardening Wine caps are also known as the "garden giant," as some of these burgundy-colored mushrooms can be massive. Growth occurs at a remarkable rate, from adolescence to delectable maturity in the span of a single day.

Purchase sawdust spawn from a reputable supplier. Choose locations with partial shade that are least likely to be disturbed to establish culture beds. We have wine caps popping up under apple trees, along the edge of the raspberry patch, between asparagus rows, even tucked among perennial herbs. Anyplace you'd want to see wood chip mulch . . . as that's the growing medium of choice. Beds can be started on open ground or cardboard sheet mulching. Lay down a layer of coarse hardwood chips (alder works especially great) a couple inches thick and then pepper with spawn. Breaking apart clumps of sawdust interlaced with white mycelium is fine as this will only stimulate greater hyphal response throughout the substrate. Add a top layer of chips to the same thickness, cover with straw, and moisten thoroughly. Weekly watering will be important during dry spells, but otherwise wine caps will be under way.

Culture beds started in spring will yield a first round of mushrooms by early fall. Fruiting occurs in three to six months depending on temperature and location. You can expect 20 pounds (9 kg) from 1 cubic yard of wood chips. Fall-inoculated beds kick in with a late-spring flush, then again in fall. Harvest when caps are just beginning to open, at the midbutton stage, for peak flavor and texture. Renewing patches annually with fresh chips keeps production coming year after year.

Europeans have long grown wine caps with corn. The traditional method involved inoculating straw and corn stubble, plowing this into the field, and then planting the corn. Mushrooms emerge at the base of corn plants while building soil at the same time. I accomplish much the same using compost rich in woodsy organic matter to prepare corn ground. Rows are made in winterkilled oat stubble with the furrow attachment on the wheel hoe; corn plugs go in one-two-three; and then I "borrow" handfuls of wood chips from established culture beds to place in the ground hither and thither down the planted row.

venison stew left to simmer on the wood cookstove as the days grow shorter.

Identifying edible mushrooms starts to get tricky really quick. Field guides with numerous photos go a long way toward showing the important indicators, but what really counts, in order to be sure, are spore print patterns and understanding fungal anatomy in every respect. Join a local mushroom club and go out foraging with folks who really know. The ten thousand mushroom species to be considered on the North American continent alone (and that's a conservative estimate) include saprotrophic fungi as well as these mycorrhizal kin. The following guides are good places to start to learn more:

Edible and Medicinal Mushrooms of New England and Eastern Canada by David Spahr

Field Guide to the Wild Mushrooms of Pennsylvania and the Mid-Atlantic by Bill Russell

Mushrooms of West Virginia and the Central Appalachians by William Roody

Mushrooms of the Midwest by Michael Kuo

Mushrooms of the Rocky Mountain Region by Vera Stucky Evenson

Mushrooms of the Pacific Northwest by Joe Ammirati and Steve Trudell

Mushrooms of the Redwood Coast by Noah Siegel and Christian Schwarz

CHAPTER 8

Soil Redemption Song

But my hand was made strong
By the 'and of the Almighty.
We forward in this generation
Triumphantly.

BOB MARLEY, LYRICS FROM "REDEMPTION SONG"

The long-term storage of carbon in the soil depends on the ability of healthy soil to form aggregates. That's achieved by mycorrhizal fungi and mycorrhizal plants. Hyphae and roots, joined as one. Each and every mycorrhiza pulsating with nutrient flow, making our lives possible. Tomorrow can be redeemed by honoring this fundamental earth pact.

We talked about how the planetary membrane of fungi protects life's sacred trust at the start of all this. That pattern, beginning with the plant cell membrane, gets repeated again and again in the physical world, including with our own body when considered in its proper context, as a community of organisms. Cohesion lies in the team approach, whether we are talking about plant ecosystems or colonized leaf surfaces or rhizosphere dynamics . . .

or indeed ourselves. This is a core teaching of the fungi. Hypha joins with hypha to form a mycelium that in turn brings a multitude of plants into a support network. Other mycelia bring other networks on board. Passage plants and bridge trees link these in turn, and so we arrive at the overarching common mycorrhizal network. Balanced nutrition and green immune function follow from there. Individuals ebb and flow, but the beat that sustains never falters.

Interestingly, some researchers choose to speak about all this in terms of warfare and competition. That mycorrhizal fungi are only beneficial up to a point. That these fungi can just as well cross the line and start taking more than will be offered back to the plant host in return. This alone changes the relationship to being parasitic in such minds. The symbiosis becomes selfishness. Even the word *host* implies a certain judgment assigned to what's taking place in the biological ego realm. In a similar fashion, many agronomists see plants as being in competition for moisture and nutrients with other plants . . . and so recommend applying herbicide to the surrounding ground in order to secure the niche of the one preferred. Truffle growers rightfully understand that other ectomycorrhizal fungi can claim tree roots for their own, thus diminishing the place at the head of the table for the truffle fungi. Yet all this is a human perspective conjured in part by the way we see ourselves. Fungi are not at "war" with other fungi. Plant communities don't evolve on the basis of superheroes defeating villains with death ray guns.

The underlying intelligence of the underground economy lies in collectively meeting the needs of the whole. We know as herbalists that the organosulfur compounds in garlic can help lower high blood pressure and that these same compounds in another body system can improve sluggish circulation. By the same token, fatty acids in holistic sprays work to deter fungal pathogens while simultaneously supporting beneficial organisms, including mycorrhizal fungi. Paradox dwells in the fact that colonized plants send significant carbon sugars to roots to fuel the soil engine — while those same plants have even more carbon for healthy growth thanks to higher rates of photosynthesis than those found in noncolonized plants. Yet what dichotomy!

We've spoken here of enzyme synthesis in plants as good for the individual and bidirectional streaming of protoplasm by the mycorrhizal fungi as benefiting the entire plant community. These are sides of the same coin, in truth. The intelligent plant diet requires that "someone" be discerning, and no one is better positioned than mycorrhizal fungi to sort out who's next in line when it comes to delivery of essential nutrients. An individual plant that takes in a surplus of nitrogen, as so readily happens in NPK-driven agriculture, becomes far more susceptible to disease.[1] The fungal network tempers excess by divvying up nutrient distribution across a wider profile of plants. Balanced

availability of trace mineral cofactors in turn drives molecular synthesis. Proteins are secured within and attached to the phospholipid bilayer. Wherever the leaf. This in turn notches up metabolism in the plant in question, making it less palatable to pests and biotrophic fungal pathogens. Yet when a single caterpillar takes that first chomp — or that errant disease spore lands from the sky and germinates — other plants are soon put on alert to prepare. You want health assurance? Have I got an assurance policy for you, my friend.

Mycorrhizal fungi are in the midst of every deal made for trace minerals and other nutrients. Every signal sent via the hyphal internet involves fungal messaging. The mere presence of fungi within the root triggers a chemical response that in turn induces resistance mechanisms throughout the plant. Plants in league with arbuscular-type fungi are constantly being gifted with biological reserves in lipid form to further robust metabolism as a result of hyphal lysis taking place within the root cortex. Even solid granite yields to the mineral weathering of fungal reach. Of all the ties that bind, mycorrhizal fungi have secured a forever place in keeping this green planet alive and on an even keel. How sweet it is that plants hooked up with these fungi back in the day!

Aldo Leopold reminded us that land health is "the capacity of the land for self-renewal." Civilizations have come and gone precisely for overlooking the importance of healthy soil. This current moment in human history only seems to be the more calamitous because the consequences of disrespecting natural ecosystems have taken on global proportions. We rightfully discuss *carbon-this* and *methane-that* in making the case to end the use of fossil fuels as soon as possible. The other part of addressing severe climate change is to work with soil fungi to restore healthy soil function.

Green vibrancy goes hand in hand with mycorrhizal connection.

and 1843. He collected 41 soil samples from farms there in order to figure out what makes certain systems more productive than others. The baseline chosen was soil organic matter. The top ten samples showed 11–37 percent soil organic matter; the bottom ten showed 2–5 percent. An eighteenfold difference in water-holding capacity was noted from one end of this continuum to the other. Today all soils in Australia fall in the 2–5 percent range. We are talking about some mind-blowing numbers from the not-so-distant past. The same scenario undoubtedly applies across the North American continent (despite our apparent lack of visiting soil scientists from Poland), as many soils today fall into this same low range. Such is the result of centuries of misguided agriculture. More to the point, we can now begin to appreciate what's possible.

Humanity needs to grasp how to make soil this good again and do so quickly. Lucky for us, the ones with such profound knowledge have stayed the course. No one knew about mycorrhizal fungi in the beginning of this *Homo sapiens* push toward depletion. Few marveled at the elegance of the soil aggregate and its unequivocal tie to planetary stability. The whole carbon-farming framework we urgently need today would have received at best an empty stare in return at the time the first plow turned the very first furrow. Perhaps it would help if we got a bit more exacting about how to think about carbon.

Carbon sequestration is a hot topic, and for good reason. More CO_2 in the sky explains in large part why our planet is warming, and dangerously so. Proposed solutions range from yet-to-be-discovered technology to government policies that put caps on carbon emissions. Requiring corporate users of fossil fuels to offset the cost of inevitable emissions by paying for green investment is another idea.[2] Averting planetary disaster will

One core aspect of the nondisturbance principle is surely doable: Don't Screw Up. We've walked through a number of such scenarios, but the one to really drive home here at the finish is to keep the ground in cover. Soils wash or blow away without plant biomass on top. Mycorrhizal fungi carry forward best when living plant partners are on the scene. Bare soil means no photosynthesis and very little biological activity. What little soil carbon has been brought about by a short-lived annual crop in the growing season soon oxidizes and goes airborne without mycorrhizae to hold down the fort. In this situation, as Christine Jones has pointed out countless times, everything becomes dysfunctional. Soil aggregates deteriorate. Water-holding capacity is reduced. Nutrients say "bye-bye."

A Polish soil scientist named Sir Paul Edmund Strzelicki visited southeast Australia between 1839 require a major reduction in the burning of oil, natural gas, and coal. A society-wide carbon tax may well be the most straightforward strategy for motivating people to reduce greenhouse gas emissions. Reducing the source of the problem, however, doesn't go far enough.

The hubbub around the recent climate treaty signed in Paris (whoop-dee-doo) introduced more of us to positive approaches for dealing with increasing carbon levels in our atmosphere. This brings the conversation back around to the role of mycorrhizae in building long-lasting soil. Had only the negotiators taken notice![3] Earlier that same year, France's Ministry of Agriculture, Agrifood, and Forestry introduced a proposal called the "4 per 1,000" initiative. This voluntary action plan focuses entirely on improving organic matter content and promoting carbon sequestration in soil through a transition to agroecology, agroforestry, and conservation management. The idea being that a "0.4 percent annual growth rate of the soil carbon stock would make it possible to stop the present increase in atmospheric CO_2." Emerging soil science estimates confirm that "we" could store 50–75 percent of current global carbon emissions in the soil. Aside from the arrogance in that last statement, reasonable people are at last talking about implementing some right-on non-disturbance techniques in agriculture and forestry.

How about that, say the fungi, *they speak-a my language.*[4]

Well, sort of. The notion that carbon simply needs to be locked away in the soil is where we come up short. Let us not forget the humic mysteries. Smaller and smaller bits of microbial matter protected from further degradation are the source of stable carbon stored in the humus fraction of the soil. Yet it is the flow of carbon through soils — rather than its absolute sequestration — that holds the keys

to healthy soil function and sustainable land use systems. Gaia lives and breathes just as we do.[5] The processes of photosynthesis, respiration, assimilation, mineralization, and oxidation are dynamic and ongoing. Mycorrhizal fungi use carbon to trade with bacteria for nutrient acquisition and to further hyphal growth. Carbon provided by plants sustains the soil food web. These same cycles hold sway in the oceans, where phytoplankton produce half of the world's oxygen and sustain the aquatic food web. The planetary carbon cycle requires the movement of carbon to keep life engaged.

That portion of carbon kept in reserve in the form of a multitude of decomposing microbes has a breath of its own as well. Such carbon sinks as found in humus and the ocean floor are vital to maintaining planetary equilibrium yet are equally dynamic, only now moving along on a far slower conveyor of time. And here's where we as humans come closer to our actual role in these states of affairs. Effective stewardship of the land involves keeping the organic matter flowing, as it's only through actively recirculating carbon that we honor fungal ways and plant wisdom. This means growing healthy crops, grazing pastures rationally, and managing forests with respect for the generations to come.

We build our lives on ecology when working in Nature's economy. The self-seeking mind-set where accumulation and consumption rule the day must be put behind us. Making such a shift requires that we see the world as sacred and valuable in its own right. Awareness of ecological fundamentals is absolutely critical. We became disassociated from all that truly sustains us when fossil fuels allowed us to forget natural systems. Now we have little choice but to revive those meaningful connections to the land and every other species with which we share this earth. The question is, can enough of us change in order to save the day? As Wes Jackson

Psychic Mycelia

We all have experiences where everything seems connected. The moment at hand transforms into a living, vibrant now. Coincidence can be expected. We zoom from seeing the microcosm to the whole of the cosmos — and then back — all in the blink of an eye. We straddle the mystery only to stand in awe of how much there is to this current life. Every breath manifests creation. Gratitude graces all.

Sharing this particular notion begins with a phone call from an orchard friend who is checking in, but mostly wanting to discuss protocols for cedar apple rust. I've just written about biological reserves in describing healthy plant metabolism the day before. Turns out Maury had heard John Kempf speak at the Practical Farmers of Iowa conference that weekend and so is buzzing about enzyme cofactors and resistance metabolites. The conversation turns to fatty acid sprays and how they impact plant defenses. Maury brings up the ideas of a David Perlmutter and the role of fats in healthy brain function. Wait for it. Just minutes before Maury's call I had picked up a book off our kitchen table brought home by Nancy as part of her nutritional studies . . . that book was *Grain Brain*, by Dr. David Perlmutter.

Comprehension shimmers like stars in the night sky.

Life does this all the time. A new matter revealed suddenly can be found at every turn. Engaging this reality proves exhilarating. Every thought links to every other thought and thereby influences the whole. The human hypha taps into psychic mycelia without end. Pulsing, flowing, and exploring the vastness. As if our very awareness is fungal in nature. Hmmm. Sure enough. The maker made things so.

puts it, "I suspect that fully living within our finite ecosphere into the foreseeable future will be the most formidable challenge our species ever has or will ever face."[6]

Restoration of the vast areas of land that have been degraded, abandoned, and overrun by weeds offers dramatic opportunities to increase both biomass and soil carbon. Setting our sights on turning such inoperable land into economically and biologically productive carbon sinks benefits everyone. The semiarid Loess Plateau grassland soils of northern China, long poorly farmed but lately undergoing transformation back to shrubland and native grasses, has seen soil organic matter content double from around 3 percent after abandonment to about 6 percent in twenty to thirty years when allowed to recover. Holistic planned grazing ups this ante yet again by actively managing livestock to heal the land. Policies that would reinvigorate the 3.5 billion acres of grasslands in the world, many of which are degraded, need to be put at the head of the line, right up there with reforestation.

Such efforts are all about restoring photosynthetic activity so mycorrhizal fungi receive carbon. The fungi in turn utilize that carbon to build soil structure, which increases soil quality. It's the best way to circulate and thus store carbon, and it's the only arrangement big enough to offset all the fossil fuels burned over the last hundred years.

The great news is that there is enough land out there in need of regeneration for everyone to do something. All approaches have merit, adding in agroforestry and perennial cropping as well. Rich choices. Real opportunity. These are indeed exhilarating times to be alive. Capitalism's decline is a necessary adjustment. Worry not, for we still have what came before, and that's the natural world and each other. Traditional cultures had an ethical relationship with Nature that worked. The same can be true again. We are not at the end of a rope, as it's so easy to think. Humanity can yet choose to turn direction. The moment has come to leap into action with glad hearts. The seeds are germinating. The fungi are willing. And we must be, too.

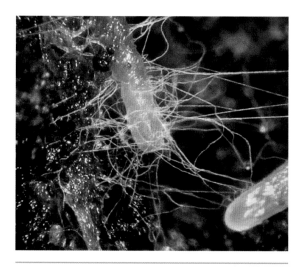

The fungus root stands revealed. Now it is up to each of us to care about mycorrhizae in the ecosystems we call home. Photo by Paula Flynn, Iowa State University.

Perhaps the real gift of the fungi and the plants isn't the "carbon solution" as much as it is showing us a way forward. That to cooperate is to find bounty for all involved. Fungal networking doesn't overlook the youngest or the least in the crowd. Older stumps are provided for as well. Most telling is that of one life giving to the next. It is written (in the book of Matthew) *for I was hungry and you gave me something to eat, I was thirsty and you gave me something to drink.* This testament, which speaks so passionately to the human heart, goes further than our species alone. Do we have ears to hear what mycorrhizae are telling us? Do we have eyes to see what tiny critters do on behalf of the whole? Beneath our feet is the teaching and the blessing.

My take on "a future and a hope" lies in tangible things my family and I can do on land we love. Each of us must embrace the good life and return to simpler ways. This connection to being, to the land, and to each other is to be celebrated. Rejoicing in this creation keeps us willing. Right now, as I write these very words, the birds are singing to greet the new day. The morning sky has begun to shimmer with all the possibility life offers anew at every turn. Ecological restoration begins when we honor the fungal component of the soil. We can do this. Generations to come need us to act now.

Nature does what needs to be done if we let her. Many people will step up to the task once provided with accurate facts and inspired by possibility. It really comes down to two simple concepts to take forward. One: Promote healthy plant metabolism as the guiding paradigm in growing anything. And two: Think and think again about ways to least disturb the soil ecosystem. Our overarching goal needs to be to amplify mycelial connection everywhere.

The fungi and the plants will sing this soil redemption song for us. As the fungi and the plants always have.

SOIL HEALTH RESOURCES

The future of agriculture and forestry lives in those who develop low-input, integrated fungal systems. Growers need what they need when they need it. From inoculum to biological equipment. All such resources expand exponentially according to the ways of fungi. I can only hope to keep up as future editions of this book roll off the presses.

MYCORRHIZAL WHEREWITHAL

Companies that offer righteous inoculum for making the world a more fungal place.

BioOrganics	www.bio-organics.com
Fungi Perfecti	www.fungi.com
Horticultural Alliance	www.horticultural alliance.com
Mycorrhizal Applications	www.mycorrhizae.com
Reforestation Technologies International	www.reforest.com
Soil Symbiotics	www.soilsymbiotics.com
Aurea (Germany)	www.mycorrhizaworld.de /hp2/Homepage.htm
Premier Tech (Canada)	www.ptagtiv.com/en

FUNGAL AND PLANT INFORMATION

Let your mind wrap around these matter-of-fact leads to greater fungal-plant awareness.

International Culture Collection of Arbuscular Mycorrhizal Fungi	www.invam.wvu.edu
International Mycorrhiza Society	www.mycorrhizas.org
Mycorrhizal Associations (Mark Brundett)	www.mycorrhizas.info
Tom Volk's Fungi	botit.botany.wisc.edu /toms_fungi
Botany Online	https://s10.lite.msu.edu /res/msu/botonl/b _online/e00/contents.htm
Plant Native	www.plantnative.com /index.htm
Plants Database (National Gardening Association)	www.garden.org/plants
Plants Database (USDA)	www.plants.usda.gov/java
State of the World's Plants	www.stateofthe worldsplants.com
Woody Plants Database	woodyplants.cals.cornell .edu/home
Rudolf Steiner Archives — The Agriculture Course (1958 Adams translation)	wn.rsarchive.org/Lectures /GA327/English/BDA 1958/Ag1958_index.html

Nondisturbance Websites

Fungal applications for those seeking practical earth truth.

Sustainable Land Management Library	www.wocat.net/en /knowledge-base /documentation-analysis /global-regional -books.html
Rhizoterra: Healthy Soil for a Healthy World	www.rhizoterra.com
Christine Jones: Amazing Carbon	www.amazingcarbon.com
Soil Carbon Coalition	www.soilcarboncoalition.org
Remineralize the Earth	www.remineralize.org
Soil Minerals	www.soilminerals.com
Advancing Eco Agriculture	www.advancingecoag.com
International Biochar Initiative	www.biochar -international.org
Terramedics Biochar Production (low-tech biochar)	www.terramedics.org /earthpitbiochar
National Center for Appropriate Technology (ATTRA) Publication List	www.attra.ncat.org /publication.html
Farm Hack	www.farmhack.org/tools
Earth Tools	www.earthtoolsbcs.com
Ecological Landscape Alliance	www.ecolandscaping.org
Harrington's Organic Land Care School	www.harringtonsorganic .com/organic-pros -program
Holistic Orchard Network	www.groworganic apples.com
Healing Harvest Forest Foundation	www.healingharvest forestfoundation.org

Ecoforestry Institute Society	www.ecoforestry.ca
Low Impact Forestry (Mitch Lansky)	www.lowimpactforestry.org
Northern Woodlands	www.northern woodlands.org
Biomimicry Institute	www.biomimicry.org
Old-Growth Forest Network	www.oldgrowthforest.net
Wildstands Alliance	www.wildstands .wordpress.com
Tree of Life Web Project	www.tolweb.org/tree /phylogeny.html
United Plant Savers	www.unitedplantsavers.org
Holistic Management	savory.global
Leopold Center for Sustainable Agriculture	www.leopold.iastate.edu
Ernst Seeds	www.ernstseed.com
North Creek Nurseries	www.northcreeknur series.com
Plant Cover Crops	www.plantcovercrops.com
Cover Crop Solutions	www.tillagebrands.com
Elk Mound Seed Company	www.elkmoundseed.com
Dawn Biologic (Duoseeder)	www.dawnbiologic.com
Cover Crop Cocktails (USDA periodic chart)	www.ars.usda.gov/Main /docs.htm?docid=20323
The Overstory (agroforestry journal)	www.agroforestry.net /the-overstory
World Agroforestry Centre	www.worldagroforestry.org
Center for Agroforestry (University of Missouri)	www.centerforagro forestry.org
Trees for the Future	www.trees.org

NOTES

CHAPTER 1:
MYCORRHIZAL ASCENDANCY

1. Fungi appear to have been aquatic at the start of the Paleozoic era (542 million years ago) according to interpretation of fossils showing spore-bearing organisms with flagella. This mobility feature would eventually be exchanged for a filamentous hyphal configuration. Fossilized hyphae and spores recovered from the Ordovician of Wisconsin (460 million years ago) resemble modern-day Glomerales, and existed at a time when the land flora likely consisted of little more than nonvascular bryophyte-like plants.

2. Seemingly immobile things *move* as the conditions around them shift. Charles Darwin's investigation into earthworms included a decades-long experiment whereby he laid down a uniform layer of chalk across a small field in 1842, and then returned in 1871 to dig a trench. During those intervening twenty-nine years, the chalk had been buried by soil and "vegetable mould" (what we now call *humus*) brought to the surface by worms. The measured depth of new soil above the chalk line was 6 inches (15 cm) in that relatively short amount of time. Wowser. The phenomenal ability of earthworms to process organic matter is something else again — but let's focus attention on the chalk layer. That appears to have "moved" downward much like plants appear to have "crawled" from sea to land. Perspective always makes for an interesting ride.

3. One cubic inch of undisturbed soil can contain as much as 8 miles of linear hyphae, according to Paul Stamets in *Mycelium Running*, based in turn on the scientific reference provided in his book. That is the basis for the foot imprint calculation. The robustness of a particular soil and climate factors

enter in here in a big way. The typical metric perspective on this number comes in more modestly at 100 meters of hyphae per cubic centimeter of prairie soil, based on another research paper dating from 1995. That can drop to a "mere" 100 centimeters in tundra soils in the Arctic. No matter how you cut it, there's an awesome highway of fungal life in the ground beneath our feet!

4. J. W. Gerdemann, "Vesicular-Arbuscular Mycorrhiza and Plant Growth," *Annual Review of Phytopathology* 6 (1968): 397–418.

5. The largest independent mycelium known occupies some 2,384 acres (965 ha) of soil in Oregon's Blue Mountains. Put another way, this humongous fungus of *Armillaria ostoyae* would encompass nearly 4 square miles (10 km²) of area. The full story can be found at http://www.scientificamerican.com/article/strange-but-true-largest-organism-is-fungus.

6. Pinaceae represents the oldest extant plant family in which symbiosis with EM fungi occurs, and fossils from this family date back to 156 million years ago. Throw in some time to develop this relationship and you arrive at scientific consensus as to when it all began.

7. J. W. G. Cairney, "Evolution of Mycorrhiza Systems," *Naturwissenschaften* 87 (2000): 467–475.

8. One Italian white truffle weighing in at 3-plus pounds (1.5 kg) sold for $330,000 at an auction in 2007.

9. Ö. Maarja et al., "Global Sampling of Plant Roots Expands the Described Molecular Diversity of Arbuscular Mycorrhizal Fungi," *Mycorrhiza* 23, no. 5 (2013): 411–430.

10. Fungal hyphae are fragile. Be it soil disturbance or nibbling by small mammals, there's an ongoing need to "repair" connections throughout the mycelium. Anastomosis provides robustness to the individual organism as well as the wider community.

11. J. P. W. Young, "Kissing Cousins: Mycorrhizal Fungi Get Together," *New Phytologist* 181 (2009): 751–753.

12. E. C. Vellinga, B. E. Wolfe, and A. Pringle, "Global Patterns of Ectomycorrhizal Introductions," *New Phytologist* 181 (2009): 960–973.

13. The notion that one fungus will knock back indigenous diversity is going to come to a head when we talk about commercial inoculum. Understanding that Nature seeks balance over "the one" will help you decide how to proceed. I suggest you peruse what Sally Smith and David Read have to say about fungus-plant specificity on pages 34–40 of *Mycorrhizal Symbiosis*, which is an in-depth discussion of this key point.

14. That unexplained *V* in the acronym for the collection comes from an earlier era of fungal thinking. This international effort began back when AM fungi were all considered to share a vesicular nature. Turns out that not all do . . . nevertheless the notation for "vesicular arbuscular fungi" continues into the modern era in assorted ways.

15. Nutrient transfer can be accomplished at numerous sites besides the arbuscular structure. Intercellular hyphae and Paris-type coils play definitive roles. The intercellular hyphae connecting everything have equal ability as well.

16. Carbohydrate currency begins with glucose, rapidly linked to become hexose, which then ships out in the form of trehalose and glycogen. A portion of these sugars will be further complexed into sugar alcohols (polyols) such as mannitol, arabitol, and erythritol for long-distance transport.

17. Extraradical hyphae extend as much as 250 meters — if stretched end-to-end — for every centimeter of colonized root. That impressive increase in absorptive surface readily snatches desired nutrients. And if you really need the English conversion to imagine the unfathomable, that's 820 feet of hyphae involved with every 0.39 inch of root.

18. One estimate suggests that the tips of ectomycorrhizal hyphae could be pushing onward into bedrock at a rate of as much as 30 micrometers per year. Nothing to write home about, but then Nature works on a much vaster time line than that of a human life span.

19. A. Rosling et al., "Vertical Distribution of Ectomycorrhizal Fungal Taxa in a Podzol Soil Profile," *New Phytologist* 159 (2003): 775–783.

20. Over time, the accumulation of soil organic matter reduces pH and inhibits nitrification. As ammonium becomes the major source of nitrogen, nitrogen replaces phosphorus as the main growth-limiting element. This favors EM plants at first. More organic matter accumulates and base depletion proceeds. Plants with ericoid mycorrhizal associations become more important because of their ability to utilize nutrients bound in acidic organic complexes. Credit goes to Dr. Read for first recognizing how nutrient availability affects plant communities through choice of mycorrhizal partners.

21. Root proliferation is apparent in a nutrient-rich patch of organic matter, whereas hyphal proliferation dominates where roots don't reach. A. Hodge and K. Storer, "Arbuscular Mycorrhiza and Nitrogen: Implications for Individual Plants through to Ecosystems," *Plant and Soil* 386 (2015): 1–19.

22. Different sources give different life spans for intraradical hyphal structures. The shortest time frame stated in one paper listed arbuscular dissolution after a mere four days. Others suggest "as much as a few weeks" when describing the time line for endomycorrhizal fungi affiliated with plants growing slowly in dappled shade.

23. Y. Kobae et al., "Lipid Droplets of Arbuscular Mycorrhizal Fungi Emerge in Concert with Arbuscule Collapse," *Plant Cell Physiology* 55 (2014): 1945–1953.

24. "The collapse and presumed digestion of arbuscules are unlikely to play a significant role in nutrient transfer from fungus to plant." Smith and Read, *Mycorrhizal Symbiosis*, 69.

Chapter 2: Healthy Plant Metabolism

1. I will relent in an endnote. Many of these phytochemicals have been called secondary metabolites simply because roles have not been found for them in primary plant functions like growth and

reproduction. Granted. Such wording allows bio-chemists to split plant metabolism into two camps. Calling the second grouping "nonessential" is where misdirection lies — reductionist assumptions too often muddle our thinking when it comes to embracing the whole of a matter.

2. One telling expression of this comes straight off the USDA website. You will find three categories through which to explore *plant health*: Plant Dis-eases, Pest Management, and Weed Management. All problem areas. Nothing about the soil food web. Nothing about mineral nutrition. Nothing about healthy plant metabolism.

3. Consider the impact of pesticides, for starters. These killer chemicals lose efficacy after a time. Pests return and the pesticide dose or the frequency of application needs to be stepped up to compensate. "Organophosphates inhibit protein synthesis . . . this is the cause of the plant's increased susceptibility, not only to sucking insects such as mites, aphids, and (so it seems) psyllids, but also to diseases, fungal and otherwise," said Chaboussou. The Green Revolution explanation of this is that the pest develops resistance. The enlightened agronomist's explanation is that the plants are weakened by repeated bouts of chemical warfare and no longer have biological support to provide balanced nutrition.

4. *Farming and Gardening for Health or Disease* became a seminal text in the organic movement soon after its publication in 1945. Sir Albert argues that industrial agriculture, emergent then and dominant today, irrevocably robs the soil of its fertility. His message was one of respecting and restoring the health of the soil in all places. Those born in these millennial times will benefit from reading such "organic classics," many of which can be found via the online library at http://www .soilandhealth.org.

5. You will learn much about trace mineral correlations and plant health from Advancing Eco Agriculture, a consulting firm started by John Kempf in 2006. Go to http://www.advanc-ingecoag.com/crop-management for specific crop recommendations featuring a wide range of mineral and biological products.

6. Theorists call such an evolutionary jump in form *symbiogenesis*. Over time each of the organisms involved loses the ability to live independently. The modern-day leaf cell with chloroplasts derived from cyanobacteria is one such result. Read more about this fascinating shift in evolutionary thinking in *Acquiring Genomes: A Theory of the Origin of Species* by Lynn Margulis and Dorion Sagan.

7. Either metal-activated enzymes have an absolute requirement for a specific metal ion, or the catalytic effect is enhanced in the presence of other metal ions.

8. A peptide can be understood to contain approxi-mately fifty or fewer amino acids. Proteins consist of one or more polypeptides arranged in a biolog-ically functional way, often bound to coenzymes and cofactors.

9. Don't be confused by fundamental definitions when considering nitrogen. Ammonium (NH_4^+) and nitrate (NO_3-) are both inorganic forms of nitrogen. Period. The basis of *organic chemistry* is the tie-in with carbon. Organic forms of nitrogen are amino acids, hydrolysates, and organic acids, which are brought about when living organisms reincorporate inorganic nitrogen.

10. Certain leaf-spotting fungi combine the two strategies. The host's immune system and cell death are actively suppressed during an initial biotrophic phase, thereby allowing invasive hyphae to spread throughout the infected plant tissue. This is fol-lowed by a necrotrophic phase during which toxins are secreted by the pathogen to induce host cell death. Such *hemibiotrophic* fungi reveal themselves fully when the lesion center falls away; shot hole fungus is the classic example.

11. Biochemists have only recently discovered that pathogenic fungi secrete "effector proteins" in an attempt to suppress immune function within the invaded plant cell. That this peptide-grade strategy finds better reception in extracellular fluid rich in amino acids is yet to be confirmed, but fits where this discussion is heading.

12. You should know that no self-respecting scientist has ever used this term. The next sentences will make clear what "haustoria" do in terms of absorbing nutrients from plants for fungal gain. Using the Heimlich maneuver (an abdominal thrust to save someone from choking) as a metaphor sort of fits, however. Squeezing out nutrients from the cell, squeezing out a blockage from the throat. And hey! A little word play now and then makes this stuff more comprehensible for us lay folk.

13. The cell membrane is a thin layer of protein and fat that surrounds the cytoplasm of a plant cell, within the more structural cell wall. The cell membrane allows some substances to pass into the cell while limiting others.

14. Disease hyphae also "take root" in the intercellular spaces between plant cells. Amino acids are absorbed by haustoria and intercellular hyphae while carbohydrate (hexose, etc.) uptake seems to be limited to haustoria.

15. A fascinating study from the 1950s revealed the proclivity of scab fungi to feed on less-than-complete proteins. The amino acids most preferred were alanine, arginine, aspartic acid, glutamic acid, glycine, and proline. "*Venturia inaequalis*: Amino Acids as Sources of Nitrogen" by R. L. Pelletier and G. W. Keitt was originally published in the *American Journal of Botany*, vol. 41, no. 4, in 1954. Research like this reveals generations of thought go into finding wholesome answers . . . eventually!

16. Phospholipid molecules have a fatty chain on one end and an acid group on the other. These organize into two layers according to polarity, with the fatty ends pointing toward the interior of the bilayer, where they are shielded from water. The "water-loving" ends face outward, where they interact with water in intracellular fluid and cytoplasm on each side of the cell membrane. Fat-soluble molecules can pass through, whereas water-soluble molecules cannot penetrate the fatty middle. Transmembrane proteins embedded throughout the bilayer serve as gatekeepers for sugars, water, and other polar molecules to pass.

17. Pathogens produce a sequence of degradation enzymes during the attack of the cuticle and cell wall. First comes cutinase, followed by cellulase, then pectinase and protease. Plants in turn have evolved a series of countermeasures to these hyphal tactics.

18. Silicic acid delivered from groundwater or through spray application of fermented plant extracts will be deposited in the walls of epidermal cells, just beneath the cuticle. As water is lost through transpiration, silicic acid concentrates and polymerizes to form "phytoliths" that buck up plant tissues and close intercellular gaps.

19. Certified organic growers applying sulfur as a "fungicide" often don't realize the primary mechanism involved. Sulfur makes the leaf surface more acidic; the spore requires more time to get its enzyme act together; hyphal penetration for necessary nutrients does not succeed prior to the leaf drying off; the spore shrivels and dies. Preventing infection through sulfur coverage works but comes with increasing biological and physiological costs when applied more than a time or two during the growing season.

20. I can certainly attest to this as a holistic apple grower. A long cloudy spell at the tail end of bloom works against orchard health in a few ways. Brix levels go down as photosynthesis falters for lack of sunshine. This should favor fruit-set pests looking for simple carbohydrates. This should favor fungal pathogens when it rains. This should hinder seed production in just-developing fruitlets. And yet somehow my trees and I get to the other side of the dark days without resorting to spray chemicals. Fat reserves see us through!

21. Polyamines help plants deal with drought, UV radiation, heavy metal toxicity, high salinity, and cold stresses by restoring the growth of cells. These everyday plant hormones metabolize from amino acids and thereby gain antioxidant properties to help plants adapt to environmental reality.

22. Each one of those sugar molecules has stored "a little bit" of the sun's energy in the form of chemical energy. There's a whole chapter of plant science concerned with ADP/ATP transporters,

proton gradients, and redox reactions that we are not exploring in this book. The physicist Richard Feynman said, "We have no knowledge of what energy [really] is. We do not have a picture that energy comes in little blobs of a definite amount." Just appreciate that you now know enough about enzymes and cofactors to be dangerous!

23. Roots take up nitrogen in the form of nitrate, ammonium, and amino acids. The nitrate portion needs to be "reduced" to ammonium in the xylem (mostly in the shoots of the plant) before continuing onward to become part of an amino acid chain. Meanwhile, the ammonium portion comes ready to go, thus saving metabolic energy. Rhizospheres with greater fungal biomass will be more acidic as a result of greater cellular respiration. More carbon dioxide leads to bicarbonate/carbonic acid generation and thus greater acidity. Nitrifying bacteria prefer a more alkaline environment. End of story.

24. Finding a suitable example of a metabolite that deters insect feeding of the beetle persuasion on a crop plant was challenging. Yet there's the dichotomy of an *alkaloid* being a *terpenoid*, which I just know some chemistry geek will eagerly call me on. Time for Google backup: "Foaming glycosides (saponins) are low molecular weight secondary plant constituents containing either a tetracyclic steroidal or a pentacyclic triterpenoid aglycone with one or more sugar chains. A broader definition would include also the steroidal alkaloid glycosides found in potatoes." The defense rests its case.

25. Azadirachtins bear a structural resemblance to the insect molting hormone ecdysone. This imparts neem with the ability to interrupt the molting cycle of insects, including the metamorphosis of pupae into adults.

26. Yeah, mate, that's a stretch. But you try carrying that tune and ending anywhere close to *Billy Shears and Sgt. Pepper's Lonely Hearts Club Band.*

27. Elaine Ingham of Soil Foodweb Inc. speaks about canopy colonization of 70 percent or more as an effective means of plant protection from pathogens. Her audio CD titled *Orchards and Vineyards* (2003) speaks to this and much more.

28. Abiotic stressors such as inorganic salts, low temperatures, and UV radiation trigger systemic acquired resistance to a degree as well.

29. A. Kessler and I. T. Baldwin, "Defensive Function of Herbivore-Induced Plant Volatile Emissions in Nature," *Science* 291 (2001): 2141–2144.

30. A brief interlude for the poets among us. Christopher Marlowe, in *Doctor Faustus* (variously dated between 1590 and 1604), referred to the beauty of Helen of Troy by these words:

> *Was this the face that launch'd a thousand ships*
> *And burnt the topless towers of Ilium?*
> *Sweet Helen, make me immortal with a kiss.*

31. The must-read paper on biologically induced systemic resistance came out two years prior to my sitting down to write this book. I finally gained research access through a web service known as DeepDyve. C. M. J. Pieterse et al., "Induced Systemic Resistance by Beneficial Microbes," *Annual Review of Phytotherapy* 52 (2014): 347–375.

32. The working definition for *elicitor* still holds, in that glycoproteins and hydrolytic enzymes of microbes do the actual eliciting of the defensive response.

33. The "biofungicide" Regalia is basically an alcohol extract of *Reynoutria sachalinensis*, otherwise known as giant knotweed. This formulated product elicits an ISR response in sprayed crops due to resveratrol as its active ingredient. The good people at Marrone Bio Innovations find it efficacious to speak in more familiar terms to farmers and thus the fungicidal construct on the product label.

34. Many commercially available insecticides are actually synthetic analogues of *pyrethrins*, called *pyrethroids*, including the insecticides permethrin and cypermethrin. What a difference a few ending letters make! Stick with the natural and only use judiciously.

35. Concentrated limonene is thought to cause an increase in the spontaneous activity of sensory nerves in insects, resulting in twitching, lack of coordination, and convulsions. Massive overstimulation of motor nerves leads to rapid knockdown paralysis. This essential oil is typically used in flea powders.

36. Biosynthesis pathways are many. Phenylpropanoid metabolism, for instance, "generates an enormous array of secondary metabolites based on the few intermediates of the shikimate pathway as the core unit. The resulting hydroxycinnamic acids and esters are amplified in several cascades by a combination of reductases, oxygenases, and transferases to result in an organ and developmentally specific pattern of metabolites, characteristic for each plant species." If you're anything like me, though, it's easy to kick back and forget all that completely.

37. Let this be the healing gift everyone reading these words checks out further. Turmeric contains curcumin, which becomes bioavailable when served in a dish with black pepper, all of which carries onward with the soluble fats in the ghee. This yellow spice has health benefits out the gazoo. Learn more at http://authoritynutrition.com/top -10-evidence-based-health-benefits-of-turmeric.

38. Nutrition science is just starting to catch on to this fundamental earth truth. "Plants' own struggle for survival — against pathogens and grazers, heat and drought — is conveyed to us, benefitting our health," according to Moises Velasquez-Manoff. Read more about mitohormesis in his excellent article published in the July 2014 edition of *Nautilus* magazine, available online at http://nautil.us /issue/15/turbulence/fruits-and-vegetables -are-trying-to-kill-you.

CHAPTER 3: UNDERGROUND ECONOMY

1. The practical takeaway to be noted about this "three-way mutualism" between AM fungi, rhizo-bacteria, and roots is that legume plantings benefit when both mycorrhizal and rhizobial inoculums are applied simultaneously to seed. Take a deep dive into all this by reading the following paper. S. Nadeem et al., "The Role of Mycorrhizae and Plant Growth Promoting Rhizobacteria (PGPR) in Improving Crop Productivity under Stressful Environments," *Biotechnology Advances* 32 (2014): 429–448.

2. EM fungi seem to have particular influence on protozoan populations in the mycorrhizosphere.

Paxillus involutus seems to reduce the density of protozoa, whereas mycorrhizae formed with *Suillus bovinus* and *Lactarius rufus* have the opposite effect. This correlation plays out in an unexpected way up above when we put common names to these fungal species. Poison Pax mushrooms are potentially fatal, whereas Red Hot Milk Caps and Jersey Cow mushrooms can be eaten safely.

3. The state motto of New Hampshire (where I live) is *Live Free or Die*. It's not exactly the easiest syntax to apply to fungi . . . but hey! A guy does what a guy does.

4. The term *saprophyte* actually applies to more than parasitically inclined fungal species. Many bacteria and protozoa, animals such as dung beetles and vultures, and a few unusual plants (including several orchids) would be deemed saprophytic for drawing sustenance from living plants as well as dead matter.

5. Armillaria root rot is a fungal root rot caused by several different members of the genus *Armillaria*. The rhizomorphs of such fungi penetrate living roots, then develop into a white mat of mycelium between the wood and the bark. Colonization gradually extends up to the root collar, eventually killing the tree outright. This major forest disease particularly affects young conifers, such as those planted during reforestation efforts or in Christmas tree plantations.

6. You can read more about how "plants take on fungal tenants on demand" at https://www.mpg .de/10390194/plants-symbiosis-phosphate. What makes this all the more poignant is that a relative of this fungus, *Colletotrichum graminicola*, is noted for causing anthracnose leaf blight and stalk rot of corn, a disease of worldwide importance.

7. Similar jobs could be contracted out to the true professionals of mycelial design. Researchers used slime molds to evaluate the Japanese rail system. Oat flakes were placed on a glass plate to reflect key transportation hubs around Tokyo. Initially, the slime mold dispersed evenly in exploring its new territory. The slime mold then began to refine its pattern, strengthening the ties between oat flakes

while other links gradually fell to the wayside. One day later, the slime mold had constructed a brilliant network of interconnected nutrient-ferrying tubes strikingly similar to the actual rail system. See http://www.wired.com/2010/01/slime-mold-grows-network-just-like-tokyo-rail-system.

8. Researchers are getting better and better at looking at whole systems in analyzing mycorrhizal networks originating from different plant species. This paper takes into account the interplay of nine different AM fungal species in linking together separately yet ultimately as one: N. Cândido et al., "Vegetative Compatibility and Anastomosis Formation within and among Individual Germlings of Tropical Isolates of Arbuscular Mycorrhizal Fungi (Glomeromycota)," *Mycorrhiza* 23, no. 4 (2013): 325–331.

9. I spent two fascinating hours with Kris Nichols in her research facility at the Rodale Institute in Pennsylvania talking about mycorrhizal everything. Know this: A guy takes notes as fast as he is able while at the same time trying to get his head around some rather complex organic chemistry and thermodynamic principles. Any slight deviation of the story line from actuality would be entirely my responsibility at this point.

10. Check out the work of forest ecologist Suzanne Simard of the University of British Columbia as to how trees communicate and care for the next generation. This article explores the differences in fungal prospects for nursery-grown seedlings versus field germinants: B. Marcus and S. Simard, "Mycorrhizal Networks Affect Ectomycorrhizal Fungal Community Similarity between Conspecific Trees and Seedlings," *Mycorrhiza* 22, no. 4 (2012): 317–326.

11. "In the days when the *information superhighway* is the buzz phrase, we would do well to look at our inventive fungal predecessors who, for four hundred million years, have already been leading a communication network of life on land." So wrote Lynn Margulis in the foreword to *Hypersea* by Mark and Dianna McMenamin more than twenty years ago. Put that in your Google and smoke it.

12. Stefano Muscano, director of the International Laboratory of Plant Neurobiology, pursues such leads at the University of Florence in Italy. Science-based research does a far better job of eliminating the sorts of variables that too often make possible desired conclusions. That buried pipe with water flowing through, for instance, is dry as a bone on the outside . . . lending credence to the argument that sound vibrations alone are redirecting root growth.

13. One such demonstration of network signaling was done with isolated tomato plants. Y. Y. Song et al., "Hijacking Common Mycorrhizal Networks for Herbivore-Induced Defense Signal Transfer between Tomato Plants," *Scientific Reports* 4, no. 3915 (2014).

14. Z. Babikova et al., "Underground Signals Carried through Common Mycelial Networks Warn Neighboring Plants of Aphid Attack," *Ecology Letters* 16, no. 7 (2013): 835–843.

15. D. Johnson and L. Gilbert, "Interplant Signaling through Hyphal Networks," *New Phytologist* 205, no. 4 (2015): 1448–1453.

16. Granted — species assemblages will always be in flux to one degree or another. Glaciers come, sea levels rise, atmospheric composition changes, fire leaves its mark. The term *climax community* can never represent a final point in the grand scope of time.

CHAPTER 4: PROVISIONING THE MYCORRHIZOSPHERE

1. Not all soil scientists are convinced that glomalin-related soil protein is derived solely from the hyphae of AM fungi. The complex structure first described as glomalin may in fact be an artifact arising from the very stringent soil extraction treatments used to obtain it. Debates aside, we can say that something in soil coming from somewhere contains a helluva lot of carbon and helps bind soil particles together. Which is good enough for a guy like me to proceed.

2. Sara Wright, along with scientists from the University of California–Riverside and Stanford

University, conducted a three-year study on semi-arid shrub land and a six-year study on grasslands in San Diego County, California, using outdoor chambers with controlled CO_2 levels to measure mycorrhizal response.

3. Glomalin for the most part is hydrophobic, thus repelling water. A hydrophilic side is exhibited as well when glomalin acts as a coating for mycorrhizal hyphae and spores. These opposing aspects are explained by the folding nature of the molecule itself: One side is sticky; the other side helps spread clay particles apart. The brilliance of Nature shows when the same molecule can show resilience in both aqueous and dry environs.

4. Lignin can make up 20–35 percent of wood, while cellulose can make up 40–50 percent and hemicellulose can make up 25–40 percent, depending upon tree species.

5. The realm of wood-rotting fungi includes the polypores that colonize the heartwood of a living tree, producing bracket-shaped fruiting bodies. Dry rot fungi cause widespread structural damage to wood-framed houses; soft rot fungi act more often on unpainted boards and driftwood. Facultative saprophytes spread from tree to tree causing root rot, yet also require deadwood to produce fruiting bodies. None of these, however, are among the cast of unruly characters with the ability to mineralize lignin.

6. Brown rots only marginally alter guaiacyl, the predominant lignin found in coniferous trees, yet in that process release certain phenols and aliphatic compounds (including benzene and napthalene) that can prove allelopathic to deciduous species. The tannin-rich bark of softwoods contributes to this suppressive effect as well.

7. White rot fungi produce numerous enzymes (including lignin peroxidases, manganese peroxidases, and laccases) that allow these saprotrophs to break down and utilize organic substrates in both guaiacyl and syringyl lignin. Lignin decomposition requires the production of oxidative enzymes that use either oxygen or peroxide to degrade the lignin polymer with free-radical mechanisms.

8. The recalcitrant litter of the boreal forest reaches its climax as a peat ecosystem. High acidity, cool temperatures, and poorly drained forest conditions contribute to this lockup of carbon as surface lignin and hemicelluloses accumulate. Climate change will lead to large increases in carbon emissions from soils at northern latitudes as permafrost soils warm.

9. That research resulted in the publication of several papers. You will find a posted copy of "Regenerating Soils with Ramial Chipped Wood" by Céline Caron, Gilles Lemieux, and Lionel Lachance at http://www.dirtdoctor.com/organic-research-page /Regenerating-Soils-with-Ramial-Chipped-Wood _vq4462.htm.

10. Both omega-3 and omega-6 are essential fatty acids that the body cannot synthesize. Such fatty acids must be obtained from the foods we eat. Both were present in our ancestors' diet in roughly equal amounts. Modern foods provide an overabundance of omega-6 fatty acids, however, chiefly from refined vegetable oils used in fried foods and packaged snacks. Omega-3s, on the other hand, are now relatively rare in American diets. Oily-fleshed, cold-water fish such as wild salmon, sardines, herring, anchovies, and sablefish are preferred species from which to obtain this essential fatty acid. Walnuts, flaxseed, eggs from chickens raised on pasture, and soy products such as tofu and tempeh contain omega-3s as well. Some people choose a fish oil supplement in lieu of these whole foods.

11. You deviant bunch of scoundrels!

12. Read this sentence again . . . and then keep healthy leaf and bark tissues in mind when we talk about fatty acid sprays for the garden and orchard. *May your protective plant layers always be lipid-rich* sounds like a righteous kind of Irish blessing to me.

13. The other class of lipid molecules, based on a branched five-carbon structure called isoprene, was first identified via steam distillation of plant materials. The resulting essential oils are often fragrant and thus used in aromatherapy and perfumes. A wide variety of structural offshoots are possible with isoprene, leading to a diverse set of compounds, including terpenes, such as

β-carotene, pinene (turpentine), and carvone (oil of spearmint); and steroids, such as testosterone, cholesterol, and estrogen.

14. A likely hypothesis is that the fungus, which lacks ability to synthesize saturated fatty acids, acquires palmitic acid from plants. The large amount of fatty acids typically found in AM fungi could be the result of a very efficient lipid transfer mechanism between the two symbionts. Furthermore, the proteins normally associated with plant defense reactions produced during hyphal penetration could be involved in this synthesis of palmitic acid from plant sugars.

15. Saponification is a process that produces soap, usually from fats and lye. Triglycerides make up the bulk of animal fats and vegetable oils. These lipids can be converted to soap with an alkaline base (such as lye), which releases the range of fatty acids contained within.

16. The database making up lipid profiles of organisms is rather extensive. Microbial ID can identify over 160 unique fatty acids (and related compounds) in a sample, including saturated fatty acids, monounsaturated fatty acids, polyunsaturated fatty acids, branched-chain fatty acids, and hydroxy fatty acids.

17. And neither should we be, for that matter, but that's another subject entirely. The right fats make all the difference. See "The Skinny on Fats" by Sally Fallon and Mary Enid posted on the Weston A. Price Foundation website at http://www.westonaprice.org/know-your-fats/the-skinny-on-fats.

18. Citrex was being marketed as a "vigorescent" with fungicidal attributes for crops such as mangoes and papaya. Such was the language of the time . . . and it would take me a few years to catch on to the actual ways that fatty acids work to support competitive colonization and help boost immune function in plants. Currently, grapefruit seed extracts for horticultural purpose only seem to be available in places like South Africa, Brazil, and perhaps Germany.

19. Fatty acid levels in neem, like any plant source, can vary from season to season, site to site. Test results from a current batch indicate palmitic acid at 18.1

percent, stearic acid at 17.5 percent, oleic acid at 46.6 percent, linoleic acid at 17.5 percent, with traces of several others.

20. Ahimsa Organics in Minnesota imports the highest-quality neem oil available, and this from a fair-trade source. Check http://www.neemresource.com for pricing information and more.

21. Typical fish hydrolysate breakdown of fat content reveals levels of combined palmitic and stearic acids at 41.5 percent, linoleic acid and other polyunsaturated fats at 17 percent, and oleic acid at 41.5 percent. The polyunsaturated fats include those desirable omega-3s at 2 percent, as the alpha-linolenic acid portion.

22. Chlorine levels are quite high when fish emulsion is boiled down to a 50 percent concentration. All the heat-sensitive vitamins, amino acids, growth hormones, and enzymes are destroyed. Less ethical companies add enzymes back into the product so they can call it a hydrolyzed process, but technically it is no more than a hydrolyzed emulsion. One good way to determine the difference is by looking at the NPK analysis: Fish naturally contains approximately 2–3 percent nitrogen. If a company is selling fish fertilizer with a higher analysis than that on the nitrogen number, it has been boiled down or evaporated or is just plain dubious.

23. CLA (conjugated linoleic acid) is a type of fat associated with a wide variety of human health benefits, including immune and inflammatory system support, improved bone mass, improved blood sugar regulation, reduced body fat, reduced risk of heart attack, and maintenance of lean body mass. Levels of CLA in grass-fed organic milk are as much as 62 percent higher than in grain-fed conventional milk.

24. Google Books to the rescue! A search preview of *Fatty Acids in Foods and Their Health Implications* makes this clear in the chapter on fatty acids in fermented food products.

25. Cut the neem oil rate by a third in the holistic core recipe, substituting karanja oil in its stead. This synergistic approach definitely improves plant health in my ever-subjective view.

26. When a triglyceride-rich oil is poured into water, the hydrophilic ends of the fatty acids interact with the water surface while the hydrophobic tails "point" straight into the air. Think about olive oil floating on top of balsamic vinaigrette . . . who knew? And thus we shake the salad dressing to better disperse those hydrophobic tails.

27. I have enough trouble properly skimming the fat from broth when making chicken soup. I can only imagine what would happen to my orchard sprayer if I then added that extracted chicken fat to the holistic core recipe!

28. Emphasis on the word *mixed*! A proper compost pile consists of alternating layers of "brown" (carbon-rich) and "green" (nitrogen-rich) organic matter. Specific tips for making fungal compost are explained ahead in chapter 6.

29. Tainio Biologicals offers forty species of micro-organisms in working with formulators. Blends are typically prepared with a generalist approach (a bit of everything) or a targeted approach (cellulose digestion, phosphorus solubilization, nitrogen fixation, chitinase activity, salt stress, and so on).

30. Several strains of *Trichoderma* are promoted as biocontrol agents against fungal diseases of plants. Endophytic advantages here include antibiosis, parasitism, inducing host-plant resistance, and competitive colonization. Most biocontrol agents are from the species *T. harzianum, T. viride,* and *T. hamatum.* These fungi generally grow on established root surfaces in their natural habitat, and so protect from root pathogens in particular.

31. *Trichoderma* attacks mycorrhizae in plate assays, granted. Results in field situations vary according to which species of fungi are being inoculated at the same time. Life is complicated that way. Ultimately, an assortment of AM fungi working in tandem with *Trichoderma* ward off root diseases in perennial plantings quite effectively.

32. Every ten thousand years or so massive glaciers come down from the north to grind exposed bedrock and thereby release a fresh round of minerals. Similarly, rising oceans leave behind perfectly balanced deposits of sea minerals when climates shift yet again. *The Survival of Civilization* written by John Hamaker and Don Weaver in 1982 launched the Remineralize the Earth movement. Joanna Campe and friends in turn have made many people aware of the use of rock dusts to help restore planetary health.

33. To be fair about this . . . these particular recommendations were taking into account soils that had been killed by glyphosate (at extreme doses) with no piles of waiting compost in sight. Nor was cover cropping under way from the previous season due to a circumstantial time line. Shipping liquid products accounts for more than 10 percent of that stated cost, this by freight delivery of drums on pallets. The even higher cost of shipping product in 5-gallon containers makes restoration budgets like this daunting on a small scale as well.

34. Modern industrial farming insists upon *high yields or bust.* Suffice it to say, I follow the recommendations of Dr. William Albrecht, an agronomist at the University of Missouri, who taught that balance is more important. The brief discussion that follows will promote "base-saturation theory" as opposed to the "sufficiency method" used by many state laboratories to maximize yields regardless of qualitative costs.

35. Mineral balance will be reflected in a rising cation exchange capacity whereby soils can hold greater amounts of exchangeable nutrients. Interestingly, as base saturation levels come into alignment, the pH will be where it needs to be.

36. Mycorrhizal fungi occur in all terrestrial environments where their host plants grow. This includes soils with varying pH, from as low as 3 to as high as 9.5. However, at the extremes of this range, the occurrence in plants becomes more limited. Mycorrhizae are extremely rare or only rudimentarily developed in calcium-rich soils such as limestone beech forests, for instance.

37. The irony here is that orchardists often use excessive amounts of copper as a mineral fungicide to combat disease. I have seen very little fire blight or brown rot here in northern New Hampshire

in the past twenty-five years and thus have very rarely applied "delayed dormant copper" for those purposes. This scenario has begun to shift with ever-warmer weather. Still, there's a balance point between *enough* and *too much* that needs to be respected. A ground-directed application of copper sulfate proved efficacious at the time, as soil levels of this element had dropped into the low range in an otherwise healthy scenario. I chose not to spray the branch canopy, as copper would certainly have disrupted arboreal organisms working for the good. And yes, copper sulfate could have been applied as a soil amendment in fairy dust fashion . . . whereas a dilute spray was merely a quicker way to get the job done.

38. Holistic orcharding offers alternatives to sulfur sprays as a means of thwarting disease. Decades ago I was making as many as twenty applications of sulfur to control apple scab disease. I transitioned this down to two or three applications as my understanding about natural systems grew. Finally came the season to go "cold turkey," and the rest is history. The irony here is that our orchard soils now need an occasional charge of sulfur. Gypsum (calcium sulfate) is a pH-neutral amendment that simultaneously boosts calcium levels.

39. Keep in mind that even a forest garden requires disturbance to get started. Designing a self-sustaining system is wonderful stuff, but there's no reason to be haughty about it from day one.

40. The respiratory openings on the undersides of leaves are flanked by guard cells that regulate transpiration by opening and closing the *stomata*. Micropores in the cuticle between guard cells and neighboring cells are more permeable than transcuticular pores elsewhere on the leaf surface. Stomate micropores allow the passage of metal chelates and other larger molecules, whereas transcuticular pores can only take in smaller ions. Furthermore, these nanometer-sized pores are lined with negative charges, so they are attractive to cations (ammonium, calcium, potassium, magnesium) but tend to repel anions (nitrate, phosphate, sulfate). Nutrient uptake into leaf cell cytoplasm works much the same as nutrient uptake by root cells once this passage through the cuticle has been accomplished.

41. The work of Russell Rodriguez and his team at Adaptive Symbiotic Technologies is to be celebrated. Partnering with symbiotic microorganisms to grow healthy crops in the face of climate change is right on.

CHAPTER 5:
FUNGAL ACCRUAL

1. Is it possible to introduce a "wayward" mycorrhizal species that might displace native fungi? The answer is yes, but it's an involved and evolving field of knowledge. Natives will maintain the status quo if introduced species lack long-term competitive abilities. Localized strains of AM fungi are more obliging than EM fungi with provincial ties to indigenous tree species. M. W. Schwartz et al., "The Promise and the Potential Consequences of the Global Transport of Mycorrhizal Fungal Inoculum," *Ecology Letters* 9, no. 5 (2006).

2. Once mycorrhizal fungi disappear from a site, reestablishment can take a long time. Soil in the blast zone associated with the eruption of Mount St. Helens back in 1980 was superheated to the point that most mycorrhizal fungi were wiped out. Vegetation has slowly been coming back, yet it struggles in those areas hardest hit. Researchers identified a twenty-year lag time on average before spore density returned to past norms. V. H. Dale, F. J. Swanson, and C. M. Crisafulli (eds.), *Ecological Responses to the 1980 Eruption of Mount St. Helens* (New York: Springer Science+Business Media, 2005).

3. Polysaccharides and fatty acids might indeed boost nutrient resources for germinating spores and vagrant hyphae on the rootward journey. Products such as Pro Vita from Soil Symbiotics do up levels of effective colonization, allowing use of fewer propagules to the same or greater effect. Spray applications of blackstrap molasses, liquid fish, and extracted humic acids to the soil surface offer a similar edge for home cooks.

4. Welcome to the world of exactitude! The spore diameter range of numerous endomycorrhizal species can be found at the International Culture Collection of (Vesicular) Arbuscular Mycorrhizal Fungi website at http://invam.wvu.edu.

5. Mesh standards differ in the English-speaking part of the world. A #50 mesh refers to the number of openings across one linear inch of screen. Such a fine-sized screen filters out all particles greater than 297 micrometers in size. Metric-oriented people would call this a 300-micron mesh.

6. Spore size of *Gigaspora margarita* ranges from 260 to 400 μm, with a mean of 321 μm.

7. Sunlight has a cumulatively negative effect on endomycorrhizal spore viability with increasing exposure times. LebanonTurf in Pennsylvania shares a number of useful mycorrhizal insights like this on its website. Check out http://www.lebanon turf.com/education/bilogicalplanttreatments /mycorrhizalfungi for the light exposure article and more.

8. Sometimes the most unexpected gems come up on an internet search. Ted St. John put together a chart of propagule counts based on soil quality and what that means in regard to average spacing between potential spores on the theoretical acre. See his full write-up on "The Instant Expert Guide to Mycorrhiza" at https://www.parco1.com/text /mycorrhiza/Mycorrhiza%20Primer.pdf.

9. Some companies present mycorrhizal counts in terms of so many propagules per gram. On that basis, a hundred or more — 132 to be exact — is essentially on the order of the pound concentration for endomycorrhizal fungi stated here. Still other companies tell you there are a minimum of so many spores per cubic centimeter (cc) . . . which equates to a fifth of a teaspoon . . . but correlating volume measure any further has to account for actual product weight. That varies with the medium chosen to retain propagules. Growers need to ask such formulators how many cc's per pound in order to compare back with the per-pound basis.

10. I asked my editor to consider letting me present this book with metric measurements alone.

Pushing the American psyche into the twenty-first century seems more than overdue. Yet noooo, Ben made it clear that the time for that hasn't arrived. So I've been converting most every number with units accordingly, trying to play nice. Now there's this *pounds per acre* bit wanting to become *grams per hectare* or some such bloody thing. And I can do it but you know what? It's going to stay hidden here in a damn endnote: A metric grower will need 67.2 kilograms of inoculum per hectare to achieve this recommendation.

11. The California Department of Transportation established a standard for restoration projects that seems to have become the guiding principle behind mycorrhizal rates. "Endomycorrhizal inoculum shall be applied at the rate of 8,900,000 live prop-agules per hectare (3,600,000 live propagules per acre), based on the guarantee of the supplier or the analysis returned by an independent laboratory." Thanks, road guys.

12. *Agrobacterium rhizogenes* injects its plasmid DNA into the genes of plant hosts to cause "hairy root" disease. Similarly, a slurry of these bacteria swiped across the revealed cells of sliced carrots will induce the hairy root phenotype. The resulting adventitious roots are subsequently introduced to sterilized spores of specific AM fungi, resulting in spore counts of five thousand or more from coaster-sized petri dishes to use as starter inoculum.

13. The aeroponic method works best with endomy-corrhizal species that produce abundant spores within the roots as opposed to further away on extraradical hyphae. Hydroponic-based systems are limited that way. *Glomus intraradices* was given that name because its spores are indeed generated within the root cortex. Fast turnaround time combined with the fact that this species is such an all-around generalist explains why *G. intraradices* is so widely promoted by single-strain manufacturers.

14. Infectivity rates based on initial propagule counts typ-ically test out more like 50–60 percent, once beyond the "root fragment window" of a granular product. Plenty of bang for the buck in year two. Frugality becomes more doubtful further down the line.

15. *Rhizopogon* and *Laccaria* species top the charts at over 60 percent infectivity after several years of aging, whereas *Scleroderma* species and *Pisolithus tinctorius* test out at 10 percent infectivity at best. Any statement about "average infectivity" needs to account for species mix, age of spores, and the habitat ahead. K. Nara, "Spores of Ectomycorrhizal Fungi: Ecological Strategies for Germination and Dormancy," *New Phytologist* 181 (2009): 245–248.

16. Spore stratification makes intuitive sense. When I saw this tidbit dangling yet again in the following paper, I knew it had to be shared. I. Louis and G. Lim, "Effect of Storage of Inoculum on Spore Germination of a Tropical Isolate of *Glomus clarum*," *Mycologia* 80 (1988): 157–161.

17. Ents are the ancient shepherds of the forest who closely resemble trees in J. R. R. Tolkien's fantasy world Middle-earth. They talk slowly and move slowly, with a sense of time more suited to trees than short-lived mortals. Ents were part of the Song of Creation, taught to speak by the elves. Picturing that we can still go forth in the spirit of Fangorn and Fimbrethil to nurture the land and speak for the trees pleases me.

18. Hyphae from other roots have a limited capacity to grow, and will die within a few days if they do not encounter a susceptible root nearby.

19. Plant list after plant list states that pome and stone fruits affiliate solely with endomycorrhizal fungi. This may well be true, but then "The Uncertainty Principle" unveiled in chapter 6 (see page 155) suggests otherwise. Equally important in this assessment are the other plants and assorted fungi that will be joining in an orchard-wide mycorrhizal network.

20. Adding manure compost or soluble NPK fertilizer to these sacks of grass would be a mistake. High fertility makes roots less dependent on mycorrhizal deals for phosphorus and nitrogen. Fewer active fungi means less spore production.

21. I. R. Coelho et al., "Optimization of the Production of Mycorrhizal Inoculum on Substrate with Organic Fertilizer," *Brazilian Journal of Microbiology* 45 (2014): 1173–1178.

22. See M. Lohman, C. Ziegler-Ulsh, and D. Douds, "A Complete How-To: On-Farm AM Fungus Inoculum Production," Rodale Institute, December 8, 2010, at http://rodaleinstitute.org/a-complete-how-to-on-farm-am-fungus-inoculum-production for full details of the mycorrhizal inoculum system developed by David Douds.

23. Mycorrhizal plants transfer fifteen times as much carbon into soil as nonmycorrhizal plants, according to Christine Jones of Amazing Carbon.

24. S. Waksman, *Humus: Origin, Chemical Composition and Importance in Nature* (Baltimore, MD: Williams and Wilkins, 1936), 62.

25. The paper that called all this humic confusion to my attention is deserving of a read. M. Kleber and J. Lehmann, "The Contentious Nature of Soil Organic Matter," *Nature* 528 (2015): 60–68.

26. Over a given year, under average conditions, 60 to 70 percent of the carbon contained in organic residues added to soil is lost as carbon dioxide. Another 5 to 10 percent is assimilated into the organisms that decomposed the organic residues, and the rest becomes "new" humus. Small fractions of this will became stable humus and pass the test of time. See the guide "Sustainable Soil Management" via the provided link for the National Center for Appropriate Technology in this book's resources.

27. The Soil Carbon Coalition does similar work in North America, which is where I came across the Christine Jones paper shared here. See http://soilcarboncoalition.org/files/JONES-Carbon-that-counts-20Mar11.pdf.

28. Different sources of what's essentially compressed humus have varied properties, depending on whether clay-based or carbonaceous shale. The marketing pitch accordingly comes down to humic and fulvic acid content based on familiar lingo. The negatively charged organic acids released from the oxidation of humates reduce the tendency of valuable minerals to electrostatically attach to soil particles. This effectively increases the cation exchange capacity of soil.

29. Those who assume "buried wood" equals "buried carbon" are mistaken. Carbon dioxide is generated

in the decomposition process as organic matter oxidizes and microorganisms breathe. The net carbon gain as exhibited by rich humus by the end of a hugelkultur run still comes down to the trade between actively growing plants and AM fungi. Soil aggregation is the means to the end every time.

30. Adding fresh manure while building the pile matrix will also help with nitrogen tie-up when using mostly newly cut wood in a hugelkultur project. I'm also serious about using urine . . . though you can be more subtle about this and *pee in a bucket* and then subsequently sprinkle some "yellow water" onto a settling pile. Similarly, if you use wood chips as part of the organic matter mix to cover the mound, providing a charge of nitrogen to any first-year plantings is an equally good idea. This can be as simple as several shovelfuls of compost beneath each squash or pumpkin vine.

31. Be aware that a continuous log dam buried under topsoil will literally blow out at select pressure points during a downpour. Allowing overflow gaps in the design makes "swale hugelkultur" a safer bet. I'd even suggest a counterintuitive approach: laying logs into the swale itself and chipping smaller debris atop that to create yet another riff on Fukuoka's idea. Only now the rubble drain runs across contour, with enriched seepage into the planting zone on the downhill side, with silt sediments over time creating a proper hugelkultur dynamic.

32. Unraveling a round bale of hay is a slick means of covering the edges of a lengthier pile quickly. Any extra soil or compost can then be pushed from above to anchor this bottom pitch and subsequently augment both hay and wood decomposition.

33. Tall grasses burn hot and quick, yet close to the ground, where air is excluded, grasses will not burn completely and thus form char. Scientists recognize that the mollisols of the midwestern prairie states contain charcoal that is structurally comparable to the terra preta soils, and far more abundant than previously thought.

34. Jane Johnson et al., "Biomass-Bioenergy Crops in the United States: A Changing Paradigm," *The Americas Journal of Plant Science and Biotechnology* (2007) as found on the web in ARS-USDA User Files.

35. Unlike a "biomass plant," which takes the firing process all the way to mineralized ash, a "biochar plant" roasts the woodsy material with little oxygen. That provides significant heat, while at the same time trapping the greenhouse gases as part of the biochar. Hospitals, schools, and small-scale manufacturers that invest in such systems can contribute to sustainable agriculture in a significant way and launch a profitable side business to cover the costs.

36. Biochar pellets recovered from the on-farm system used to propagate *Glomus intraradices* alone exhibited twenty-four propagules g^{-1} fresh weight. D. Douds et al., "Pelletized Biochar as a Carrier for AM Fungi in the On-Farm System of Inoculum Production in Compost and Vermiculite Mixtures," *Compost Science & Utilization* 22, no. 4 (2104). Published online at http://handle.nal.usda.gov/10113/59997.

37. High phosphorus applications markedly decrease arbuscular fungal biomass. The mechanisms hindering fungal development are mediated entirely by internal plant-derived signals. S. E. Smith et al., "Roles of Arbuscular Mycorrhizas in Plant Phosphorous Nutrition," *Plant Physiology* 156 (2011): 1051.

38. The strength of the extract results in different numbers. Mehlich and Bray readings are provided for acid and neutral soils whereas the Olsen readings are applicable to alkaline soils.

39. What's known as the Morgan Soil Test provides readings for phosphorus and potassium that reflect the available portion of these two elements. Converting these particular phosphorus and potassium readings into a pounds-per-acre basis (multiply ppm times two) are a useful means for evaluating nutrient-density potential. I explore this thoroughly in *The Holistic Orchard* if you're interested . . . but now we're heading back to biological prospecting.

40. Soils low in organic matter may contain only 3 percent of their total phosphorus in the organic form, but high-organic-matter soils may contain 50 percent or more of their total phosphorus content in

the organic form. A newly planted tree with a developing fungal root system initially lacks the reach to access the fungal duff zone envisioned over time.

CHAPTER 6:
PRACTICAL NONDISTURBANCE TECHNIQUES

1. "Antidepressant Microbes in Soil: How Dirt Makes You Happy" by Bonnie Grant can be found at http://www.gardeningknowhow.com/garden -how-to/soil-fertilizers/antidepressant-microbes -soil.htm.

2. Flavonoids act as signaling compounds in the AM symbiosis. The flavonoid content of root exudates is stimulated by the cytokinin hormone found in reconstituted kelp. The resulting chemical signal is one means by which mycorrhizal hyphae find a desirable root host. Additionally, gibberellic acid is considered a germination signal for seeds. All this from a seaweed soak!

3. T. E. Cheeke, T. N. Rosenstiel, and M. B. Cruzan, "Evidence of Reduced Arbuscular Mycorrhizal Fungal Colonization in Multiple Lines of Bt Maize," *American Journal of Botany* 99, no. 4 (2012): 700–707.

4. Far too many gardeners take it for granted that corn seed must be treated with a fungicide to prevent seedling diseases in cold, wet ground. Currently, most corn seed is treated with fludioxonil, a broad-spectrum contact fungicide. Metalaxyl is often included to control pythium blight. That locally systemic addition prevents mycorrhizal colonization of the corn root system for three weeks, if not more. *There is no joy in Mudville* here, folks.

5. No one needs to take this little math anecdote about loss of organic matter any further. Too many factors go into determining the actual SOM cost brought on by a single pass of a rotary tiller. It may be true that in some soils the loss of SOM is as little as 1 percent. This would be significantly more percentage-wise for a degraded soil with already low levels of organic matter. Yet the fact that I once got it into my head that this single percent loss represented a "full point" off recorded soil test amounts is funny only because it helped me view tillage as a gardening practice to mostly avoid.

6. The increase in soil temperatures under black plastic mulch has been found to shift the soil organism community toward a bacterially dominated soil ecosystem. The Rodale Institute has published a twenty-four-page guide to help growers consider more earth-friendly alternatives to the plastic mulch craze, which is available at http://rodale institute.org/beyond-black-plastic.

7. Take a break for a well-deserved book plug. Jean-Martin Fortier's *The Market Gardener* picks up the trail blazed by Eliot Coleman in *The New Organic Grower*. Every organic entrepreneur will want to be aware of the innovative ideas shared in this indispensable guide to biologically intensive microfarming.

8. R. Valeria, B. Paolo, G. Manuela, and H. Marcel, "Mycorrhizal Fungi Suppress Aggressive Agricultural Weeds," *Plant and Soil* 333 (2010): 7–20.

9. An inner ring of peastone comes highly recommended right up against the trunk. Organic mulch keeps bark a tad too moist. The peastone will be useful in suppressing weedy growth close in and makes a nice base for a permanent vole guard.

10. A lipid polymer layer impregnated with waxes protects all surfaces on vascular plants. Cuticular wax primarily blocks nonstomatal water loss. Other functions of the cuticle include deterring insects and pathogens, obstructing UV penetration, and keeping surfaces clean of spores and airborne particulates.

11. The year is 2009. The distribution of a disease organism with such devastating impact via national retail chains is domestic bioterrorism of the stupidest sort. It's worth noting that transplants sold in local garden centers that had obtained tomato seedlings from local sources were free of late blight.

12. It's almost impossible to halt the further development of late blight once 5–10 percent of the foliage appears infected. A long streak of very dry and hot weather (both day and night) might temporarily stop an epidemic. Stem infections are very resistant to drying, however, and will sporulate when sufficient moisture becomes available.

13. This would have been an extreme sanitation measure in the good ol' days. Sexual reproduction of *Phytophthora infestan* requires two different mating types. Up until the early 1990s, the late blight pathogen could only reproduce asexually via sporangia in North America because all strains were of the same mating type. Both mating types are now present in the United States and Canada, however, so sexual reproduction of late blight is a real possibility. And that's the reason exposed plant parts must go or be thoroughly composted in a hot pile.

14. Seeding a cover crop into corn and soybean fields by airplane prior to harvest speaks to the grain seed's ability to take hold on the soil surface. Successful "aerial seeding" works best when it rains soon after broadcasting and the crop about to be harvested provides shading. A few extra pounds of seed per acre than would be required if drilling or incorporating make up for any seed that doesn't quite make it to the ground.

15. Check out the tool ideas shared by the Farm Hack online community at http://farmhack.org/tools. Type "culticycle" into the site search engine and invent away!

16. Throw in a smattering of German or Japanese millet (and even annual medic) to keep mycorrhizal affiliation to the fore when using buckwheat as a summer cover crop. Buckwheat alone will weaken fungal connections. That's not an issue when using buckwheat in a smother rotation to beat back rhizome-type grasses early on in a new garden. The plants listed here will "mow-kill" and so not interfere with garlic plans.

17. The perceptive reader will note a coming together here of two techniques. Woodsy compost for orchard use involves far more setting time to develop its full fungal component. The last-minute addition of ramial chipped wood in creating "garlic compost" is more for developing fungal connections over the winter period ahead. The thin layer transition outlined by the Quebec team doing the original research into ramial chipped wood was all about building long-term fertility tied to the seasonal cycles of soil organisms. This biological approach to garlic growing accomplishes that very thing, only with the crop already in place.

18. You don't want to overdo nitrogen, as this can undermine mycorrhizal affiliation in the early stages of spring growth. One season I applied liquid fish to my garlic beds at a very high rate in hopes of making up for lack of enough compost when prepping the beds the previous fall. That seemingly innocent act led to a withdrawal of mycorrhizal fungi that otherwise would have scavenged nitrogen for the inoculated garlic plants. A very warm spring with little rain followed . . . allowing common *Fusarium* fungi to exploit an unexpected niche. This resulted in basal root rot devastating nearly half of that year's planting. Robust soil dynamics were essentially undone by a guy not yet fully versed in biological underpinnings. An application of *Trichoderma* (beneficial endophytic fungi) at the first sign of yellowing leaves would have shifted this situation back to an even keel. Remember all this when "your turn" comes.

19. Tree roots in a poorly dug hole in clay soil fare worst of all. Root growth slows when roots reach the undisturbed site soil beyond the backfill zone. Lower oxygen levels in compacted soil and the "glaze effect" (from shoveling straight down through clay to create the planting hole itself) contribute equally here.

20. Most trees do not have taproots, in truth. Some trees, such as oaks and pines, will develop deep roots directly beneath the trunk in sandy, well-drained soils. These vertically directed roots will not go down all that far, however, when the water table is close to the soil surface or when the soil is compacted. Only species in the walnut family (which includes pecan and hickory) are "carrot-rooted" with very few laterals until firmly established.

21. Placing the root ball on undug ground prevents the tree from sinking and tilting as the soil settles. Backfill should not be placed up against the trunk but rather taper down to grade starting from several inches out. The root ball typically "sags" an inch or more over time, and then all will be nearer to grade level.

22. Mycorrhizal Root Dip from BioOrganics, for one, contains nine types of AM spores as well as five EM spores. It also contains a horticultural hydrogel (which likely cannot be used by a certified organic operation) that forms a root-clinging slurry when mixed into water. A granular inoculum works just as well if you're only planting a few trees.

23. Such may not be the case for trees kept too long in containers, where circling roots often interweave just below the surface. Current research finds that the old standard of slitting the revealed root ball on four sides does not adequately deal with circling roots. Shaving off the outer inch or so of the root system with a knife is now recommended. *Ouch.* This encourages otherwise pot-bound roots to grow outward, however, so do it.

24. Read up on "haphazard mulching" in *The Apple Grower*. Being less than uniform about such things is simply another form of diversity. Similarly, depending on your goals, it's perfectly acceptable that taprooted weeds and wildflowers grow back in aging mulch. The real goal here is not the look above but rather the fungal element below.

25. Mineral amendments (including lime) can be worked in accord with the expanding radius approach as well. I often do this with orchard trees to spread out upfront costs.

26. I have a slight peeve with many of the fruit guild designs encouraged in permaculture courses. The concept of stacked functions is certainly right on, but, frankly, you don't put massive plants such as comfrey or lovage a mere foot or two (30–60 cm) out from the trunk. Young fruit trees need more "personal space" to get a proper start. Airflow and sunlight around the trunk zone have value.

27. Use ramial chipped wood as a bedding base in animal stalls instead. The wiry portion of hay, autumn leaves, straw, chopped cornstalks, and even "strewing herbs" can then be spread on top as needed. These are the fixings of good compost in the making, whereas the woody portion of conifers in the form of sawdust or shavings is a different carbon scenario entirely.

28. Mellowing time also suits ramial chipped wood garnered from species with allelopathic reputations of their own. Black walnut is the classic example. All parts of this tree contain juglone, which is used by the walnut tree to suppress nearby competition. The volatile oils found in eucalyptus species work similarly.

29. In that same vein, when transplanting perennials into an already mulched bed, it's a good idea to pull back the mulch somewhat to provide "elbow room" for new plantings to get roots into ground warmed by sunshine.

30. The active decomposition zone is best viewed as an intermediate veneer between the rhizosphere underneath and hyphal rumblings in the mulch layer above. Conditions at the soil-mulch interface are more acidic as a result of fungal activity. This does not mean that woodsy mulches make the soil as a whole more acidic. No scientific research supports this. One study found neither pine bark nor pine needles had any effect on soil pH outside the interface zone. Similarly, ongoing wood chip mulch does not tie up nitrogen because decomposition takes place at this interface with the soil, and nitrogen is rereleased into the soil at about the same rate at which is tied up. Reality is in flux from one state to the next.

31. B. Appleton et al., "Mycorrhizal Fungal Inoculation of Established Street Trees," *Journal of Arboriculture* 29, no. 2 (2003): 107–110.

32. Subsoilers are implements used to loosen and break up soils to double the typical depths that a disk harrow reaches. The purpose here is to lift and aerate the soil while limiting soil disturbance deeper down, thus minimizing oxidation of organic matter. A hefty tractor is required to pull even a single shank. A pass made down each envisioned row can help fruit trees establish sooner. So-called *keyline plowing* involves a series of subsoiling shanks pulled along the contours of the terrain to improve moisture availability.

33. Glyphosate is the active ingredient in Roundup, the ubiquitous herbicide that Monsanto has "gifted" upon this world of woe. This soil poison

works by tying up essential nutrients and thereby shutting down plant immune systems. Plants become completely vulnerable to pathogens and die. Glyphosate has been shown to be toxic to rhizobia, the bacterium that fixes nitrogen in the soil. Glyphosate significantly limits root symbiosis with arbuscular mycorrhizal fungi by decreasing AM biomass, vesicles, and propagules. Glyphosate has recently been indicted as a probable human carcinogen by the World Health Organization. Turning this evil tide requires the types of practices spoken about in this book: microbial inoculation, cover cropping, humates. All applied in good time.

34. The worst scenario I've ever encountered played out this way. A good man with dreams of starting up a cidery ordered a couple thousand trees for spring planting a year out. But a first land deal fell through, and it wasn't until late March that papers were signed on another property. The ground on this abandoned farm had been in grapes a dozen years back when a disgruntled employee deliberately overdosed thirty-some acres of trellised vineyard with Roundup herbicide. Yeah. Flash forward to readying this very ground to plant cider apple trees ASAP. The crew faced a tangled jungle of sizable Russian olive overgrown into the once taut trellis wires. Patches of mugwort and goldenrod were about the only other plants present in this ghost landscape. And this was the very week that trees should have been going in! Kudos to Pete and the boys for achieving the impossible. Still, despite significant inputs and mycorrhizal inoculation, rest assured it's going to take a couple of years for these courageous trees to come up to speed.

35. Case in point, Japanese knotweed. *Fallopia japonica* is quite possibly the best nectary going in early autumn for native pollinators and beneficial wasps. Its virtues include high resveratrol levels in root rhizomes . . . which in turn provide herbalists with a key remedy for Lyme disease. Think remedy yet again as an inducer of systemic resistance in fruit trees, as explained in chapter 2. The hollow stems of knotweed can be used to build up populations of solitary nesting bees. You can eat the young shoots like asparagus. And if you talk nice to this plant (and mow along the edges), it will indeed stay relatively put.

36. Cation balance is about far more than adjusting pH. Getting fertility ratios right among calcium, magnesium, and potassium is incredibly important to future fruit quality. This perspective leads to choosing the appropriate kind of lime — calcitic or dolomitic — or even gypsum if pH is already ballpark. You can read more about soil chemistry considerations for orchards from an eco-ag perspective in both *The Apple Grower* and *The Holistic Orchard*.

37. No need to go overboard with supplemental nitrogen, however, since nitrogen-fixing legumes will be included in any good cover crop plan. Still, a smattering of alfalfa meal or soybean meal can make a difference if soils are truly depleted. Ground sprays that include liquid fish will help as well once cover crops have germinated. The primary goal here is to grow organic matter, which in turn feeds soil organisms, which in turn feed healthy fruit trees. Supplemental nitrogen for cover crops should be viewed as a bulking catalyst to get the healing process under way.

38. Pardon me, but it's time for a car analogy! This legume step is akin to shifting into second gear on a fungal transmission. We let off the clutch slowly as the biology starts rolling in first gear, gain some speed with *Trifolium pratense*, and will be ramming it home as fruit trees take root. Tom and Ray Magliozzi would be proud.

39. Humanity owes a lot to Professor Gilles Lemieux (and Céline Caron and the others!) who essentially explained the evolution of fertile soil in a forest ecosystem, and then applied this knowledge to agriculture.

40. This equals about 80 cubic yards per acre, or 2 cubic yards per 1,000 square feet. The larger metric conversion works out to about 150 cubic meters per hectare.

41. Certain bacterial species enter a dormant period as *cysts* when faced with unfavorable environmental conditions like cold. Cell walls thicken, metabolic processes slow down, and the bacterium ceases

all activities like feeding and locomotion. Other bacterial species enter dormancy as highly resistant structures called *endospores* that form within the original cell. These can be considered much the same for our purposes here.

42. Bigger wood can go in multiple directions. First and foremost is into the woodstove for heat. Larger limbs can be part of a hugelkultur project, subsequently buried for long-term fertility banking. Relatively straight limbs 6 to 8 feet long stacked "Lincoln-log-style" create functional (removable) compost bins. Our farm outhouse features curvaceous apple slabs framing the doorway to create a womblike entrance. Whatever the means, redirecting bigger wood keeps the carbon-to-nitrogen ratio of ramial wood chips in the fungal sweet spot.

43. I will issue a SPECULATION ALERT on this statement. Nowhere have I seen research into the rate that soluble lignins become more recalcitrant or if this matters one iota. This echoes that all-too-familiar question of if ramial wood chips should first be aged . . . fresh seems best . . . but we humans don't really know when fungi digest what for the greater good, do we?

44. One of my apple buddies had the seemingly brilliant idea to take the prunings from his standard-sized trees to create an impenetrable hedge wherever deer regularly accessed the orchard to nibble on younger trees. Time passed and the surrounding branch mounds grew ever larger. It takes three to four years for black rot to establish and subsequently make itself known by sending spores out by the tens of thousands around bloom time. My friend spent several growing seasons dealing with frogeye leaf spot (primarily with copper sprays) before *one and one made two* . . . and now the deer can come and go as they please.

45. Summertime surgery in the form of pruning out fire blight strikes will be very important in preventing canker formation in ever-larger limbs. Learn about the "ugly stub technique" in my book *The Apple Grower* or from extension literature as a means of holding further infection at bay. Proper pruning cuts can then be made in colder winter months.

46. As do peaches and cherries and plums. The dynamics of flower formation, however, change . . . and so to keep the story line that follows spot-on . . . this up-front bit of distinguishing between the pome fruits and the stone fruits is in order.

47. Tree hardiness winds its way down from terminal buds, through that recent shoot growth, into the limb, and eventually round the trunk. Apple and pear trees are not pruned during this phase in order to avoid stimulating a growth response that would interfere with hardening off. This "senescence period" for fruit trees across temperate zones starts at some point in mid-August and lasts through leaf abscission, whereupon the tree comes into full dormancy.

48. Only one of every ten white feeder roots subsequently browns and goes woody, thereby extending the tree's permanent root system. The other nine slough off, thereby providing additional organic matter for microbes.

49. Orchard systems abound. Our great-grandparents often kept orchard ground cleanly cultivated to give trees priority. The infamous herbicide strip seemingly addresses this, but at unreasonable biological cost. Organic fruit trees on dwarfing root systems often are shallow-cultivated along row edges to lessen nutrient competition for those smaller root systems. A mycorrhizal approach to orcharding encompasses far broader thinking.

50. Flowering plants in a biodiverse orchard ecosystem will be somewhat in harmony with seed formation in grasses. Red clover at this fungal point in time, for instance, will have first bloom to as much as 20 percent of its flowers showing. This seed formation window occurs over a period of weeks, so don't get hung up looking for an "exact moment" to mow underneath and around orchard trees.

51. The pleasure of scything will only be understood once you have used the European rendition of this preeminent mowing tool. Simply put, a relatively straight snath makes for a straight spine. Laying down dripline mulch consists of two revolutions around each tree. First reaching under the lower branches to pull green mulch out with short

strokes, then going back around to lay even more mulch atop that. The scythe allows you to pick and choose as well. Living mulch plants such as comfrey, just coming into bloom, should be left standing for bumblebees to enjoy.

52. It's time for a primal lesson about grass. Regular mowing creates a denser carpet of grass blades, which is deemed desirable by some. Thus, *the lawn*. That shearing of the green on a regular basis produces a similar effect down below, where grass root systems are as much as twenty times denser per cubic foot of soil volume than would be found in a meadow ecosystem. All roots respire carbon dioxide. Feeder roots of fruit trees can deal with normal levels of carbon dioxide, but upon encountering "lawn levels" dive further down in search of soluble nutrients. Far better for tree health and immune function that feeder roots access the upper humus layer of the soil and partially built nutrition.

53. Some technicalities here: The apple scab fungus releases *ascospores* in spring to rekindle its disease cycle. Cedar apple rust bounces back and forth from eastern red cedar trees to apple trees in the form of *basidiospores*. Brown rot on stone fruit overwinters as mycelium on twig cankers and mummified fruit. These become sources of *conidia* that infect blossoms, which in turn sets up a second stage of actual fruit rot closer to harvesttime. *Monilinia fructicola* causes "brown rot" on peaches and the like . . . which may seem confusing to those of you recalling the brown rots that decay softwood. Different fungi entirely.

54. The following words come direct from a patent application found online: "Fatty acid compositions are useful for disrupting existing microflora (pathogens and saprophytes), making it more likely that a biocontrol organism will successfully colonize the surface if it is applied at the same time or soon after the fatty acid composition." And thus was born the fatty acid knockdown as first expressed in *The Holistic Orchard*. Use of the FAK in late fall or prior to budbreak in early spring gives growers a biological alternative to spraying copper. And what a great acronym to boot!

55. Liquid fish provides a nitrogen boost in addition to fatty acids and other good enzymes. That charge of foliar nitrogen is inappropriate come August and through the harvest season when buds need to "tuck into dormancy" and thus gain winter hardiness. Come leaf fall it is another matter entirely. Fish rates in the holistic fall spray are doubled to boost nitrogen levels and thus encourage greater leaf decomposition.

56. Azadirachtin is found only in neem seeds. These limonoid compounds bear a structural resemblance to the insect-molting hormone ecdysone. This imparts neem oil with the ability to interrupt molting, metamorphosis, and development of the female reproductive system. Immature insects exposed to azadirachtin (mainly by ingestion) may molt prematurely or die before they can complete a properly timed molt. Those insects that survive are likely to develop into deformed adults incapable of feeding, dispersing, or reproducing.

57. It's only fair to point out there's a "pear exception to the rule" when it comes to safe foliar rates. Pear leaves are apparently sensitive to any oil application immediately following bloom. Far better not to spray pears with neem at petal fall and the immediate weeks beyond than to see blackened leaves the next day. Dropping the neem rate to 0.25 percent during this time period would be worth trialing where situations (like pear blister mite) warrant.

58. There are three components of any arbuscular root system: the root itself and two associated mycelial structures. Soil-based hyphae are involved in the acquisition of nutrients and search for new plant partners. Hyphae within the root itself act as intraradicle interfaces where nutrient transfers take place. Boundaries between different fungal species become hazy at this point, particularly in the intercellular spaces of the root.

59. *Frankia alni* is a species of actinomycete (filamentous bacterium) that forms nitrogen-fixing nodules on the roots of alder trees.

60. Would this allelopathic effect be applicable to fruit trees? I don't have direct experience with

eucalyptus to be able to say. Correspondingly, would EM hyphae reaching beyond the root systems of eucalyptus somehow carry this allelopathic effect along? The upshot here is that eucalyptus might not be an ideal bridge tree.

61. What fun to confound the issue even further. Should you indeed be a cabinetmaker, you probably have worked with green-hued "poplar" as a less expensive hardwood suitable for painting. This lumber term almost always refers to a specific wood species in the United States: *Liriodendron tulipifera.* Other common names for this hardwood include yellow poplar, tulipwood, or tulip poplar. The only problem with referring to this species as "poplar" is that it isn't actually even a poplar! That title properly belongs to a genus of soft-wooded trees appropriately named *Populus.*

62. There's no way to make this stretch without issuing a full-blown SPECULATION ALERT. Such a list of possibilities based on lumber designation alone might also include buckeye, pawpaw, tree of heaven, striped maple, hedge maple, swamp maple, northern catalpa, and eastern redbud.

63. Nitrogenous *green* material includes grass clippings, alfalfa meal, vegetable wastes, and urine-soaked muck from the barnyard. Leaf mold, straw, cornstalks, woodsy brush, and moderate amounts of drier stable bedding (wood shavings) are much higher in carbon than nitrogen, and thus are recommended as *brown* material.

64. This paper was originally published in 1987 in *New Phytologist,* volume 105, no. 2. Internet access is made possible through JSTOR, a niche database set up for librarians, publishers of scientific journals, and individuals.

65. Some means of aeration will help sustain bacterial action and higher temperatures within static compost piles. Perforated pipe can be installed at the base of the pile and again midway, placed horizontally and extending beyond the circumference of the pile. Bundles of coarser sticks placed to project upward at a slight angle (as the pile gets built) provide air access into the heart of a pile. The resulting chimneylike convection works best with other stick bundles placed horizontally in the "cleanout door" position.

66. Others have referred to compost as black gold — which always triggers a certain association for a guy like me. Bluegrass musicians Lester Flatt and Earl Scruggs introduced every episode of *The Beverly Hillbillies* throughout the 1960s singing "The Ballad of Jed Clampett." Listen in at YouTube (http://www.youtube.com/watch?v=0_XAPku7SgE) as you read on about fungal compost. *Oil that is, black gold, Texas tea.*

67. Much ado has been made about the allelopathic effect of softwood-derived lignin in the jaws of nonenzymatic brown rots. Polyphenols and other compounds released early on can set back deciduous species. That's reason enough not to use fresh softwood chips to mulch younger trees. Months later, however, those same softwood chips are ready for fuller decomposition by an entirely different group of fungi.

68. Ramial wood chips made from summer stock (with leaves) heat things up beyond the desired fungal influence. Nitrogen-rich chips can be used in moderation for the initial phase of composting, but should not be used in the maturation phase when microbes are all about cooling down.

69. Mycorrhizal fungi are not present in composts unless the compost contains plant roots. No roots, no inoculation. And really? Even this forest-edge advantage will be overkill for those working biologically from the very start.

70. Of course earthworms play a big role here! Along with springtails, potworms, oribatid mites, wood lice, millipedes, snails, and slugs. Breakdown of organic matter involves a diverse team of players, ending with the work of fungi and bacteria in the release of nutrients.

71. Time a fall holistic spray when approximately half of the leaves are on the ground. The tree canopy gets coverage as well, so even leaves still on the tree become "tastier" for decomposition purpose. Many objectives are achieved by stirring the biological stew by such means, as detailed in *The Holistic Orchard.*

72. Time to talk about scab sex in this quiet corner of a reference note. *Venturia* fungi have two mating types that form *pseudothecial initials* within a month after the leaves fall. Scab enters a period of dormancy in colder zones at this point. By early spring, mating has taken place in the leaves between the two mating types. (You probably think I'm making this language up, but no, this is pure science talk from the American Phytopathological Society.) Spore sacs called *pseudothecium* form as a result of this dalliance. The world turns and love goes on.

73. Hooked you in with my literary license this time! Quebec studies have shown that a post-harvest application of *Microsphaeropsis ochracea* (formulated as a conidia suspension) can be used to reduce the number of potential scab ascospores on leaf litter by 75 percent. Here's the thing. This very same biological control agent will already be at work in your orchard, provided the understory ecosystem supports a wide range of decomposing organisms. Leaf litter feeds that crowd.

74. The high rate of transpiration in wooded areas enriches the atmosphere with water vapor. This in turn creates a "condensation gradient" whereby the forest canopy sucks in moist air from the ocean. Ergo, remove the majority of trees in a region and there will be fewer sustaining rains to keep the plant-fungal cycle running smoothly. The *biotic pump* is the major driver behind bringing moisture across continents. This theory developed by Russian physicists Anastassia Makarieva and Victor Gorchakov explains many things about radical climate change.

75. G. Likens et al., "Effects of Forest Cutting and Herbicide Treatment on Nutrient Budgets in the Hubbard Brook Watershed-Ecosystem," *Ecological Monographs* 40 (1970): 23–47.

76. The percentage values that follow should be considered provisional, as the sampling intensity varied widely among biomes (fourteen tropical stands, three boreal, and only one temperate). K. K. Treseder and A. Cross, "Global Distributions of Arbuscular Mycorrhizal Fungi," *Ecosystems* 9 (2006): 305–316.

77. All numbers concerning logging system impact in this section are courtesy of the untiring work of Mitch Lansky. His book, *Low-Impact Forestry*, speaks to bringing back a human-scale approach to managing woodlots sustainably.

78. Mitch Lansky is quoted in W. Sugg, "Low Impact Forestry Gaining Ground in Maine: A Three-Part Strategy to Save and Restore Forests," *Maine Organic Farmer & Gardener*, Spring 1998.

79. I know I am working this fungal analogy to the hilt . . . yet it really applies, no? Just as hyphae redirect nutrient resources in ad hoc style by joining spontaneously (anastomosis) to form a vibrant mycorrhizal network, our lives intersect with one another all for the better when we integrate local resources fully into a rewarding economy for all.

80. Mycorrhizal Applications has done numerous inoculation trials with various crops. A spore blend of four species (*G. aggregatum*, *G. intraradices*, *G. mosseae*, *G. etunicatum*) always proves better than one. Greater biomass results when fungal diversity is at play, as reported to me by Dr. Mike Amaranthus in an April 2016 phone conversation.

81. A "mixing barrel" for making small batches of mortar works great for seed inoculation on a market garden scale. Check out the OdJob Mixer available from A. M. Leonard's Gardening Edge website.

82. Effective microbe cultures that have been activated to increase spray volume (approximately twenty-two times over) are pushed toward a dormant state once blackstrap molasses sugars are consumed and the brew becomes more acidic. This makes for a slight lag time on the leaf, especially with regard to the photosynthetic bacteria. The so-called purple haze attribute of effective microbes will best be galvanized by twelve to twenty-four hours of aeration (stir, baby, stir) prior to application. Look into *Rhodopseudomonas palustris* to learn fascinating details about this ancient bacterium and its potential in agriculture.

83. Start with forest soil (complete with mycelium) and go from there in creating an indigenous beneficial organism culture. The diverse microbe sources and provided foodstuffs in this recipe are

well thought out. See http://permaculturenews
.org/2012/02/04/how-to-prepare-a-beneficial
-microorganism-mixture.

84. USDA researcher Sara Wright found 0.7 milligram
glomalin per gram of soil in a nearby field that was
plowed each year, compared with 1.7 milligrams
per gram of soil found in a field after three years
of no-tilling. The gold standard, by comparison,
would be as much as 15 milligrams per gram of
glomalin in undisturbed grasslands.

85. Dr. Don Huber of Purdue University has said
farmers were once encouraged to rotate herbicides
because if one herbicide killed off a group of soil
microbes, they might have a chance to come back
under a different chemistry. Since Monsanto's
Roundup Ready crops hit the scene in 1996,
farmers have been slamming fields with glyphosate
every year, usually multiple applications per year.
This has eliminated the chance for soil microbes to
repopulate in any meaningful way.

86. Gabe Brown builds healthy soils on significant
acreage in North Dakota with a wide array of cover
crops. This conventional farmer came to recognize
that "roots in the ground" deserve the credit for
a good two-thirds of organic matter increase. See
"Diversity Is King" in the October 2013 edition of
Acres, USA for a fascinating read on what it means
to step back and learn from the land.

CHAPTER 7:
EDIBLE MYCORRHIZAL MUSHROOMS

1. Arora, *Mushrooms Demystified*, 35.

2. Arora, *Mushrooms Demystified*, 191.

3. The lethal webcaps, two species in the genus
Cortinarius, are two of the world's most poisonous
mushrooms. The species are the deadly webcap (*C.
rubellus*) and the fool's webcap (*C. orellanus*). The
mushrooms' characteristics are quite common,
making them difficult to identify, which often leads
to fatal poisonings.

4. Let's take this one step further. Truffle fungi produce a steroid identical to a pheromone produced
by boars (male pigs) during premating behavior.
Turns out this same hormone, androstanol, is

also secreted by male humans, though at a much
lower concentration. Some say this explains the
aphrodisiac claims made for truffles. Forget the
body cologne, guys, and truffle up instead.

CHAPTER 8:
SOIL REDEMPTION SONG

1. A flood of nutrients in soluble form undermines
the need for the plant to form mycorrhizal
affiliations in the first place. Strike one comes with
the nitrates that feed disease organisms. Strike two
is loss of fungal dominion and all the symbiosis
benefits that follow from there. Strike three is the
subsequent disconnect in obtaining the array of
trace minerals required for optimal protein synthesis. Such a plant would now be called "out of here"
if we were playing baseball. Which is a good analogy, for now the plant likely requires medication
(in the form of fungicides) to compensate for these
obvious biological and nutritional deficiencies.
Whatever is being grown loses any remaining ties
to enduring biology. The resulting crop has far less
virtue in terms of nutrient density and medicinal
oomph. All that can be said is that surely we made a
wrong turn into this endnote.

2. Getting into government policies concerning
climate change isn't really within the scope of this
book. I will go so far, however, as to share a related
bit as heard on National Public Radio this past
year. The two commentators were discussing the
15,000 miles (2,400 km) of air travel that had been
necessary to cover a particular story. The carbon
emissions from those flights could supposedly
be made up by paying an additional $50 to plant
eighty-eight tree seedlings somewhere. Such is the
rationale of "carbon offsets" in the unreal world.
And while planting trees is always good, it's fair to
say that the fungi are not at all impressed by the
need to make such flights in the first place.

3. The absence of agriculture as a component
of necessary solutions at the United Nations
Climate Change Conference (COP21) in Paris in
December 2015 disappointed many. John Roulac of
Nutiva and GMO Inside strongly argues that this

kid-glove treatment can be attributed to corporate shenanigans. Overlooking industrial, degenerative farming practices is no innocent act of omission. "It's plain to see why Monsanto and friends, via their high-level political appointees, influenced the U.S. and United Nations delegates at COP21. They eliminated agriculture and soils from the COP21 agenda and thus the final agreement — despite overwhelming evidence that soil sequestration (carbon farming) is the number one solution to stop the rise of CO_2." J. Roulac, "The Fraudulent Science at COP21 Exposed," EcoWatch, December 17, 2015, http://www.ecowatch.com/the-fraudulent-science-at-cop21-exposed-1882130411.html.

4. Please smile now and have a Vegemite sandwich.

5. In Greek mythology, Gaia (pronounced 'gaɪ.ə') was the great mother of all; the primal Greek Mother Goddess; creator and giver of birth to the earth and the starry heavens above. James Lovelock in turn formulated the *Gaia hypothesis*, which proposes that organisms interact with their inorganic surroundings on earth to form a synergistic, self-regulating, complex system that helps to maintain and perpetuate the conditions for life on the planet. This personification of the earth as a living being suits very well indeed.

6. Wes Jackson, "Nature's Way: A Path to Ecological Agriculture," Great Transition Initiative (April 2016), http:www.greattransition.org/publication/natures-way.

GLOSSARY

anastomosis. The ability of fungal hyphae of the same species to fuse together, thereby allowing nutrients and defense signals to be conveyed between plants.

actinomycetes. Gram-positive bacteria active in decomposition of organic matter that give rise to the rich earthy smell of freshly tilled soil.

angiosperms. This taxonomic class includes plants whose seeds are surrounded by an ovule (think of an apple). This group is often referred to as *deciduous hardwoods*. Angiosperms have broad leaves that usually change color and shed every autumn.

arbuscule. The nutrient-transfer structure formed by arbuscular mycorrhizae within the root cell wall yet separated from the cytoplasm by the plasma membrane.

Ascomycota. Fungi in this phylum are characterized by bearing sexual spores in a sac. *Ascomycetes* include approximately thirty-three thousand species of unicellular yeasts, green and black molds, powdery mildews, morels, cup fungi, and true truffles, in addition to most lichens and the so-called imperfect fungi.

aseptate hyphae. Fungal filaments that lack cross walls, allowing for unimpeded flow of protoplasm.

assimilation. The uptake of nutrients directly from a mineral source or via the immobilization pathway of microorganisms. The net result of depositing inorganic ions in the soil food web and in plant matter prevents otherwise soluble nutrients from leaching away.

Basidiomycota. Fungi in this phylum are characterized by basidiospores borne upon a clublike structure. Basidiomycetes include approximately 22,500 species of mushrooms, polypores, crusts, corals, clubs, basidiolichens, and gastromycetes that passively release their spores (puffballs, false truffles, stinkhorns, and birds' nests). Teliomycetes (rusts) and Ustomycetes

(smuts) are obligate parasites of insects or plants that would also be included in this phylum.

biostimulant. An organic substance that stimulates plant growth and thus benefits soil biota as well. The plant growth hormones in seaweed extracts, for instance, derive from the cytokinins found in kelp.

bridge tree. Multiple mycorrhizal affiliations by certain woodsy plants make possible the transfer of nutrients between networks of different mycorrhizal types. Deciduous species sharing AM and EM affinity are the so-called *soft hardwoods*.

cambium. A layer of meristematic plant tissue located between the inner bark and wood that produces new bark and wood cells, causing the stem or trunk to grow in diameter and forming the annual ring in trees.

cation exchange capacity. The capacity of a soil for exchange of positively charged ions (cations) between the soil and the soil solution, and thus a measure of nutrient retention capacity.

chelate. The process by which certain large molecules form multiple bonds with a micronutrient, protecting it from reacting with other elements in the nutrient solution and increasing its availability to the plant.

chitin. The structural element of fungal cell walls, composed of a modified polysaccharide that contains nitrogen.

colonization. Host plants were once said to be "infected" by mycorrhizal fungi. This sounded far too pathological and thus scientists have learned to speak in terms of colonization instead.

common mycorrhizal network. The overarching collaboration of multiple fungal networks with multiple plant species across an ecosystem.

competitive colonization. Colonization by friendly microorganisms of plant surfaces denies fungal and bacterial pathogens the chance to gain an infection foothold.

cytoplasm. The fluid that fills the cell interior, consisting primarily of water, salts, and proteins. Organelles (such as chloroplasts) are embedded in cell cytoplasm as well.

dripline. The canopy diameter of a tree carried to ground level, which roughly delineates where the mass of feeder root action takes place in the humus.

ectomycorrhizal fungi. A type of mycorrhizae that associates with conifers as well as numerous hardwoods such as oaks. The fruiting bodies of EM fungi can be counted among the wild mushrooms found on forest floors.

endomycorrhizal fungi. Arbuscular mycorrhizal fungi associate with more than 80 percent of terrestrial plants. Rock 'n' roll territory, to be sure. Crop plants depend on an AM affiliation to thrive.

endophyte. A symbiont bacterium or fungus that lives within a plant for at least part of its life cycle without causing apparent disease.

enzyme. A macromolecular catalyst that accelerates biochemical reactions.

epidermis. The outermost cellular layer that covers the whole plant structure, including roots, stem, leaves, flowers, and fruit. This single layer of epidermal cells comes closely packed with few chloroplasts if any at all.

ericoid mycorrhizae. Commonly associated with wetland or bog plants such as blueberries, cranberries, heathers, rhododendrons, and azaleas. The extraradical mycelium of ericoid mycorrhizae and its associated intracellular coils have a relatively limited nutrient reach.

extracellular digestion. The process by which microorganisms secrete enzymes in order to digest surrounding material, whether organic matter or mineral rock.

extraradical hyphae. The filamentous structures that make up the fungal body in the soil. These acquire nutrients, propagate the association, and produce spores and other structures.

facultative. Capable of but not restricted to a particular function or mode of life. *Facultative anaerobes*, for instance, can use oxygen but also have anaerobic (lacking oxygen) methods of metabolism.

flavonoids. A group of compounds containing a characteristic aromatic nucleus and widely distributed in higher plants, often as a pigment. Flavonoids play a role in immune function of plants and serve as signaling molecules to initiate mycorrhizal affiliation.

fungal duff zone. The soil ecosystem around trees deliberately managed to be fungal-friendly through provision of woodsy organic matter.

green immune function. Phytochemical plant defenses based upon resistance metabolites and essential oils.

gymnosperms. This taxonomic class includes plants whose seeds are not enclosed in an ovule (such as a pinecone). This group is often referred to as *coniferous softwoods*. Gymnosperms usually have needles that stay green throughout the year.

Hartig net. The nutrient-transfer structure formed by ectomycorrhizal fungi that weaves between the outer cortex cells in tree roots.

haustoria. The nutrient-transfer structure formed by pathogenic fungi within the leaf or stem cell wall.

humic substances. Once believed to consist of humic acid, fulvic acid, and humins . . . but now better understood as the *humic mysteries*. Only kidding! Smaller and smaller bits of microbial matter protected from further degradation are the source of stable carbon stored in the humus fraction of the soil.

humification. The conversion of fresh organic matter to more stable forms of organic matter in the soil.

humus. Stable carbon compounds that remain for a long, long, long time after many organisms have used and transformed the original organic matter. Humus is not readily decomposed because it is physically protected inside soil aggregates or tightly bound to crystalline clay particles.

hyphae. The long, branching filamentous structures of a fungus, collectively called a *mycelium*.

inoculate. To introduce fungal mycelia or spores for purposes of making plants healthier.

intercellular hyphae. Hyphae that grow between the walls of adjacent root cells. These are located in the root *apoplast* — the zone outside the cytoplasm of cells — but a secret word otherwise not appearing in this book. Please do not tell *The Illuminati* that I dared to reveal the name of this heresy in such an unexpected way.

intraradical hyphae. Hyphae that grow within the cortex of a root and develop nutrient transfer structures. These make up the fungal body inside the roots of plants.

lysis. The disintegration of a cell by rupture of the cell wall or membrane and the subsequent release of cytoplasmic compounds.

macroaggregate. Conglomerations of soil particles held together by a protein-based glue of biological origin. Soil aggregates provide refuge and food for microfauna and microflora. Mycorrhizal fungi in turn utilize lignin-derived macroaggregates as a center of operations for directing more complex forms of nutrition onward to plant partners.

mantle. A hyphal sheath formed by ectomycorrhizal fungi outside of the root surface.

mineralization. The breakdown of organic matter by microorganisms into soluble nutrients that are readily available for uptake by plants. The act of "microbe consuming microbe" lies at the heart of bioavailability of nutrients.

mycelium. The vegetative body of a fungus, consisting of a mass of branching, threadlike *hyphae*.

mycoforestry. An ecological forest management system implemented through the stewardship of mycorrhizal and saprotrophic fungi to enhance forest plant communities.

mycorrhiza. The symbiotic pairing of fungus and root. The plural *mycorrhizae* finds its place in American vernacular while *mycorrhizas* holds sway among British stalwarts.

mycorrhizal network. Considered in the singular form, this term refers to the anatomized hyphae of one mycorrhizal species that tie like-minded plants together to transport nutrients and moisture from one location to another. Bear in mind that the Mighty Mycelium has a far vaster vision in mind beyond this somewhat insular loop.

mycorrhizal root length. An estimate of the length of roots colonized by mycorrhizal fungi, further convertible to percentage colonization when divided by plant root length.

mycorrhizosphere. The region just beyond roots that includes the hyphae of mycorrhizal fungi and associated microorganisms, including the immediate soil under this direct living influence.

niche. The function or position of a species within an ecological community. A species' niche includes the physical environment to which it has adapted, as well as its role as producer and consumer of food resources.

nonhumic biomolecule. Identifiable chemical structures such as polysaccharides, glycoproteins, waxes, and lignin in the humus fraction of soils.

nutrient depletion zone. The zone around roots or fungal hyphae in which the concentration of nutrients is lower than in the bulk soil because of the uptake by said roots or hyphae.

obligate. When both species in a symbiotic relationship depend entirely on the other for survival. *Obligatory plants*, for instance, are those species that will not survive to reproductive maturity without a mycorrhizal association.

parasite. An organism that depends upon another organism for its nourishment and growth.

passage plant. Multiple mycorrhizal affiliations by one plant make possible the vascular transfer of nutrients to different networks of the same mycorrhizal type.

phytoalexin. Antimicrobial compounds that are both synthesized by and accumulated in plants after exposure to microorganisms and other elicitors.

phytoanticipins. Antimicrobial plant compounds that are produced following actual intrusion by pests and disease organisms.

phytolith. A rigid, microscopic structure made of silica deposited within different intracellular and extracellular structures of the plant.

plant sap. Fluid transported in xylem tubes or phloem cells of a plant. Xylem cells transport water and inorganic nutrients through the plant; phloem cells transport carbon sugars and other biological molecules.

propagules. An assembly of spores, root fragments, and nonassociated hyphae that carry fungal colonization forward.

protoplasm. Everything contained within the cell membrane of a fungal or plant cell, including both the cytoplasm and nuclei.

protozoa. Prior to the advent of molecular biology, flagellates and ciliates were classified together as *protozoa*. These unicellular organisms are of the kingdom Protista, the predecessors to animals, land plants, and fungi.

rhizomorphs. Mature mycelial cords are composed of wide, empty vessel hyphae surrounded by narrower sheathing hyphae, looking much like actual roots.

rhizosphere. The immediate zone around plant roots where nutrients are absorbed directly by the root and a humongous population of microbes gathers to savor carbon-rich root exudates.

root cortex. The bulk of plant tissue to be found between the epidermis and vascular core of a root. These mostly differentiated cells are responsible for the transportation of nutrients and water into the central cylinder of the root through diffusion and may also be used for energy storage in the form of starch.

saprophytic fungus. A species that lives on decaying organic matter yet also may obtain nutrients by parasitizing plant hosts.

saprotrophic fungus. A species that obtains nutrients from nonliving organic matter by absorbing soluble organic compounds. Such fungi are sometimes called *saprobes*.

sclerotia. Compact masses of hardened mycelium, containing food reserves, used by fungi as reproductive bodies to survive environmental extremes.

soil organic matter (SOM). The sum total of organic compounds in the soil, including undecayed plant and animal tissues, partial decomposition products, microbial biomass, and the humus portion.

spore. A seedlike depository of fungi containing genetic material to propagate the species either immediately or after a period of dormancy.

sporocarps. Any of several structures whose primary function is the production and release of spores. Soil particles may be included in the spore mass.

symbionts. Either of two organisms that live in symbiosis with each other.

terpenoids. Any of a large class of organic compounds that occur in all plants, including terpenes, diterpenes, and sesquiterpenes. What counts here is that terpenoids play a significant role in the immune function of plants.

vesicle. Storage structures formed by intraradical hyphae within the root cortex. These may form within or between cells. Vesicles accumulate lipids and may also function as propagules.

BIBLIOGRAPHY

Arora, David. *Mushrooms Demystified*. 2nd ed. Berkeley, CA: Ten Speed Press, 1986.

Bailey, Liberty Hyde. *The Holy Earth*. Originally published: New York: Charles Scribner's Sons, 1915. Reprint of 1943 Christian Rural Fellowship edition by Dover Publications, 2009.

Bonsall, Will. *Essential Guide to Radical, Self-Reliant Gardening*. White River Junction, VT: Chelsea Green, 2015.

Bruges, James. *The Biochar Debate*. White River Junction, VT: Chelsea Green, 2010.

Brunetti, Jerry. *The Farm as Ecosystem*. Austin, TX: Acres USA, 2014.

Cardon, Zoe G., and Julie L. Whitbeck, eds. *The Rhizosphere: An Ecological Perspective*. Waltham, MA: Academic Press, 2007.

Chaboussou, Francis. *Healthy Crops*. Charlbury, England: Jon Carpenter Publishing, 2004.

Chamovitz, Daniel. *What a Plant Knows*. New York: Scientific American, 2012.

Chevallier, Andrew. *Encyclopedia of Herbal Medicine*. New York: DK Publishing, 2000.

Cook, Langdon. *The Mushroom Hunters*. New York: Ballantine Books, 2013.

Cotter, Tradd. *Organic Mushroom Farming and Mycoremediation*. White River Junction, VT: Chelsea Green, 2014.

Crawford, Martin. *Trees for Gardens, Orchards, and Permaculture*. East Meon, Hampshire, England: Permanent Publications, 2015.

Datnoff, Lawrence E., Wade H. Elmer, and Don M. Huber. *Mineral Nutrition and Plant Disease*. St. Paul, MN: American Phytopathological Society, 2007.

Eisenstein, Charles. *Sacred Economics*. Berkeley, CA: Evolver Editions, 2011.

Erasmus, Udo. *Fats That Heal, Fats That Kill*. Burnaby, BC: Alive Books, 1986, 1993.

Eshel, Amram, and Tom Beeckman. *Plant Roots: The Hidden Half*. 4th ed. Boca Raton, FL: CRC Press, 2013.

Flack, Sarah. *The Art and Science of Grazing*. White River Junction, VT: Chelsea Green, 2016.

Fortier, Jean-Martin. *The Market Gardener*. Gabriola Island, BC: New Society Publishers, 2014.

Fukuoka, Masanobu. *Sowing Seeds in the Desert: Natural Farming, Global Restoration and Ultimate Food Security*. White River Junction, VT: Chelsea Green, 2013.

Goreau, Thomas J., Ronal W. Larson, and Joanna Campe, eds. *Geotherapy: Innovative Methods of Soil Fertility Restoration, Carbon Sequestration, and Reversing CO_2 Increase*. Boca Raton, FL: CRC Press, 2014.

Hall, Ian R., Gordon T. Brown, and Alessandra Zambonelli. *Taming the Truffle*. Portland, OR: Timber Press, 2007.

Hemingway, Toby. *Gaia's Garden: A Guide to Home-Scale Permaculture*. White River Junction, VT: Chelsea Green, 2001, revised 2009.

Jacke, Dave, and Eric Toensmeier. *Edible Forest Gardens*. White River Junction, VT: Chelsea Green, 2005.

Karban, Richard. *Plant Sensing and Communication*. Chicago: University of Chicago Press, 2015.

Kimmerer, Robin. *Braiding Sweet Grass*. Minneapolis, MN: Milkweed Editions, 2013.

Kohm, Katherine A., and Jerry F. Franklin. *Creating a Forestry for the 21st Century: The Science of Ecosystem Management*. Washington, DC: Island Press, 1997.

Korn, Larry. *One Straw Revolution: The Philosophy and Work of Masanobu Fukuoka*. White River Junction, VT: Chelsea Green, 2015.

Kourik, Robert. *Understanding Roots*. Occidental, CA: Metamorphic Press, 2015.

Lansky, Mitch, ed. *Low-Impact Forestry*. Hallowell, ME: Maine Environmental Policy Institute, 2002.

Lee-Mäder, Eric, Jennifer Hopwood, Mace Vaughan, Scott Hoffman Black, and Lora Morandin. Xerces Society Guide: *Farming with Native Beneficial Insects*. North Adams, MA: Storey Publishing, 2014.

Lengnick, Laura. *Resilient Agriculture*. Gabriola Island, British Columbia: New Society Publishers, 2015.

Lowenfels, Jeff, and Wayne Lewis. *Teaming with Microbes*. Portland, OR: Timber Press, 2010.

Mancuso, Stefano, and Alessandra Viola. *Brilliant Green*. Washington DC: Island Press, 2015.

Margulis, Lynn, and Dorion Sagan. *Acquiring Genomes: A Theory of the Origin of Species*. New York: Basic Books, 2002.

Marley, Greg. *Chanterelle Dreams, Amanita Nightmares*. White River Junction, VT: Chelsea Green, 2010.

Marschner, Petra, ed. *Marschner's Mineral Nutrition and Higher Plants*. 3rd ed. Waltham, MA: Academic Press, 2012.

Mehrotra, R. S., and Ashok Aggarwal. *Fundamentals of Plant Pathology*. New Delhi: McGraw Hill (India), 2013.

Moyer, Jeff. *Organic No-Till Farming*. Austin, TX: Acres USA, 2011.

Mudge, Ken, and Steve Gebrial. *Farming the Woods*. White River Junction, VT: Chelsea Green, 2014.

Nabhan, Gary Paul. *Growing Food in a Hotter, Drier Climate*. White River Junction, VT: Chelsea Green, 2013.

Nardi, James. *Life in the Soil*. Chicago: University of Chicago Press, 2007.

Osentowski, Jerome. *The Forest Garden Greenhouse*. White River Junction, VT: Chelsea Green, 2015.

Paul, Eldor, ed. *Soil Microbiology, Ecology, and Biochemistry*. 4th ed. Waltham, MA: Academic Press, 2015.

Perry, David A., Ram Oren, and Stephen C. Hart. *Forest Ecosystems*. 2nd ed. Baltimore, MD: John Hopkins University Press, 2008.

Pershouse, Didi. *The Ecology of Care*. Thetford Center, VT: Mycelium Books, 2015.

Peterson, Larry, Hugues Massicotte, and Lewis Melville. *Mycorrhizas: Anatomy and Cell Biology*. Ottawa, CA: NRC Research Press, 2004.

Phillips, Michael. *The Apple Grower*. Rev. ed. White River Junction, VT: Chelsea Green, 2005.

Phillips, Michael. *The Holistic Orchard*. White River Junction, VT: Chelsea Green, 2011.

Rainer, Thomas, and Claudia West. *Planting in a Post-Wild World*. Portland, OR: Timber Press, 2015.

Schwartz, Judith D. *Cows Save the Planet*. White River Junction, VT: Chelsea Green, 2013.

Shetreat-Klein, Maya. *The Dirt Cure*. New York: Atria Books, 2016.

Smith, J. Russell. *Tree Crops: A Permanent Agriculture*. Originally published: New York: Harcourt & Brace, 1929. Facsimile reprint of 1950 Devin-Adair edition by Island Press, 1987.

Smith, S. E., and D. J. Read. *Mycorrhizal Symbiosis*. 3rd ed. Waltham, MA: Academic Press, 2008.

Spahr, David. *Edible and Medicinal Mushrooms of New England and Eastern Canada*. Berkeley, CA: North Atlantic Books, 2009.

Stamets, Paul. *Growing Gourmet and Medicinal Mushrooms*. 3rd ed. Berkeley, CA: Ten Speed Press, 2000.

Stamets, Paul. *Mycelium Running*. Berkeley, CA: Ten Speed Press, 2005.

Steiner, Rudolf. *Agriculture: A Course of Lectures*. Translated by Catherine E. Creeger and Malcolm Gardner. Kimberton, PA: Bio-Dynamic Farming and Gardening Association, 1993.

Stephenson, Steven L. *The Kingdom Fungi*. Portland, OR: Timber Press, 2010.

Syltie, Paul. *How Soils Work*. Fairfax, VA: Xulon Press, 2002.

Taiz, Lincoln, and Eduardo Zeiger. *Plant Physiology*. 5th ed. Sunderland, MA: Sinauer Associates, 2010.

Toensmeier, Eric. *The Carbon Farming Solution*. White River Junction, VT: Chelsea Green, 2016.

Voisin, André. *Soil, Grass and Cancer*. Reprint. Austin, TX: Acres USA, 2003.

Walters, Charles. *Fertility from the Ocean Deep*. Austin, TX: Acres USA, 2012.

Weiseman, Wayne, Daniel Halsey, and Bryce Ruddock. *Integrated Forest Gardening*. White River Junction, VT: Chelsea Green, 2014.

White, Courtney. *Grass, Soil, Hope*. White River Junction, VT: Chelsea Green, 2014.

Wolfe, David W. *Tales from the Underground*. Cambridge, MA: Perseus Books, 2002.

Zimmer, Gary. *Advancing Biological Farming*. Austin, TX: Acres USA, 2011.

Zimmer, Gary. *The Biological Farmer*. Austin, TX: Acres USA, 2000.

INDEX

Note: Page numbers in *italics* refer to photographs and figures; page numbers followed by *t* refer to tables.

ABOUT THE AUTHOR

Frank Siteman

Michael Phillips is a farmer, writer, carpenter, orchard consultant, and speaker who lives with his wife, Nancy, and daughter, Grace, on Heartsong Farm in northern New Hampshire, where they grow apples and a variety of medicinal herbs. Michael is the author of *The Apple Grower* (Chelsea Green Publishing, 2005) and *The Holistic Orchard* (2011), and teamed up with Nancy to write *The Herbalist's Way* (2005). His Lost Nation Orchard is part of the Holistic Orchard Network, and Michael also leads the community orchard movement at www.GrowOrganicApples.com

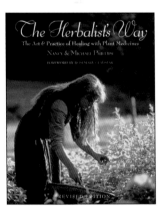